Series in
Mathematical Biology
and Medicine
Vol. 11

TOWARDS A MATHEMATICAL THEORY OF
COMPLEX BIOLOGICAL SYSTEMS

SERIES IN MATHEMATICAL BIOLOGY AND MEDICINE

Series Editors: **P. M. Auger and R. V. Jean**

Series in
Mathematical Biology
and Medicine
Vol. 11

TOWARDS A MATHEMATICAL THEORY OF COMPLEX BIOLOGICAL SYSTEMS

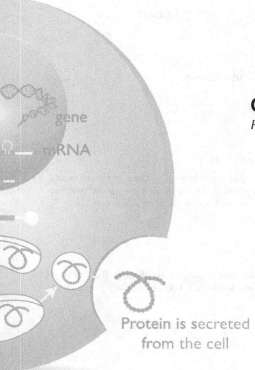

gene

mRNA

Protein is secreted
from the cell

C Bianca • N Bellomo
Politecnico di Torino, Italy

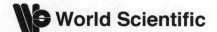

World Scientific

NEW JERSEY • LONDON • SINGAPORE • BEIJING • SHANGHAI • HONG KONG • TAIPEI • CHENNAI

Published by

World Scientific Publishing Co. Pte. Ltd.
5 Toh Tuck Link, Singapore 596224
USA office: 27 Warren Street, Suite 401-402, Hackensack, NJ 07601
UK office: 57 Shelton Street, Covent Garden, London WC2H 9HE

Library of Congress Cataloging-in-Publication Data
Bianca, Carlo.
    Towards a mathematical theory of complex biological systems / by C. Bianca and N. Bellomo.
      p. cm. -- (Series in mathematical biology and medicine ; v. 11)
    Includes bibliographical references and index.
    ISBN-13: 978-981-4340-53-3 (hardcover : alk. paper)
    ISBN-10: 981-4340-53-7 (hardcover : alk. paper)
    1. Biomathematics. 2. Biological systems--Mathematical models. I. Bellomo, N. II. Title.
    QH323.5.B4533 2011
    570.15'118--dc22

                                                                    2010043493

British Library Cataloguing-in-Publication Data
A catalogue record for this book is available from the British Library.

Printed in Singapore.

Preface

This monograph focuses on the ambitious aim of developing a mathematical theory of complex biological systems with special attention on the phenomena of ageing, degeneration, and repair of biological tissues under individual self-repair actions that may also take advantage of medical actions and therapies. The immune system plays an essential role in the competition against pathologies. Of course, the authors do not naively claim that such objective is effectively achieved. Simply some perspective ideas, and some preliminary steps, are brought to the attention of applied mathematicians.

The approach to the mathematical modeling of biological systems needs to tackle the additional difficulty generated by the peculiarity features of living matter. For instance the lack of invariance principles which are typical of systems pertaining to inert matter, the ability to express a strategy for individual well-being, heterogeneous behaviors, competition, proliferative and/or destructive actions, learning ability, evolution, and many others.

The presentation is organized in three parts with ten chapters, which have the aim of addressing applied mathematicians to research activity in the field of modeling complex biological systems viewed as living, and hence biological systems:

Part I is devoted to a phenomenological description of the aforementioned biological processes with the aim of capturing the characteristics that play a crucial role in the modeling process. A new systems biology approach is subsequently proposed, where the overall system is decomposed into subsystems on the basis of the biological functions that are expressed by each subsystem.

Part II presents the mathematical tools that are designed to achieve the aims of the monograph. These tools define a class of integro-differential equations that offer an appropriate underlying mathematical structure that can be used to derive models at the molecular and cellular scale. Moreover, mathematical methods to link the dynamics from the molecular to the cellular scale, and from this to that of tissues are studied in view of applications.

Part III deals with applications and perspectives. First, the modeling of two case studies is presented, namely the modeling of malignant keloid formation, and the derivation of chemotaxis models at the macroscopic scale from the underlying description at the cellular scale. Subsequently a critical analysis is proposed to understand how much work is still needed to achieve a mathematical theory of biological systems, and to identify the guidelines necessary to pursue this objective.

In more details, the sequential steps of the mathematical approach are as follows:

i) *Modeling biological systems* as a large, heterogeneous, multiscale complex system;

ii) Integration of the concept of *functional subsystems* to reduce the complexity of large living systems within the approach of systems biology;

iii) Development of the *mathematical kinetic and stochastic game theory for active particles*;

iv) Modeling interactions at the molecular and cellular scale and application of the theoretical approach to some case studies;

v) Development of a *multiscale analysis and modeling from genes to tissues.*

The final aim of this monograph is to provide a new conceptual background for applied mathematicians involved in the challenging research field of mathematics of living systems. The authors trust that the validity of the methodological will not be limited to the specific cases treated in the monograph, but it will provide the background for a large variety of problems generated by the interdisciplinary approach of mathematics and biology.

The authors are indebted to the Partners of the Project RESOLVE of the European Union. Their contribution to understand complex biological phenomena and to their modeling by an interdisciplinary mathematical-biological approach has been a precious gift.

Torino,   *Carlo Bianca*
September 2010   *Nicola Bellomo*

Acknowledgments

The authors acknowledge the support by the European Union FP7 Health Research Grant number FP7-HEALTH-F4-2008-202047-RESOLVE, the support by the FIRB project RBID08PP3J-Metodi matematici e relativi strumenti per la modellizzazione e la simulazione della formazione di tumori, competizione con il sistema immunitario, e conseguenti suggerimenti terapeutici, and the Compagnia di SanPaolo, Torino, Italy.

Funded by the European Commission
FP7 Health Research
Grant number FP7-HEALTH-F4-2008-202047

Contents

List of Figures

List of Tables

Chapter 1

Looking for a Mathematical Theory of Biological Systems

1.1 Introduction

This monograph tackles the ambitious aim of developing a mathematical theory of biological systems focused on ageing, degeneration, and repair of biological tissues under individual self-repair actions that may also involve advantage of medical actions and therapies. More precisely, this chapter provides a brief description of the aims, contents and organization of the monograph, organized as Lectures Notes. Specifically, it introduces the concept of mathematical theory of complex biological systems and describes the sequential steps to pursue the aforesaid objective. Of course, the authors do not naively claim that such objective is effectively achieved. Simply some perspective ideas, and some preliminary steps, are brought to the attention of applied mathematicians.

As is known, mathematical models attempt to describe, by means of equations, the dynamics in time and space of real systems that, in the case under consideration, belong to living matter. All mathematical models need to identify the parameters that appear in the equations. Once this assessment has been properly performed, the model can be used to provide an approximate description of a physical reality. In some special cases, the model can even depict emerging behaviors that are only partially shown by empirical data. Finally, it contributes to the refinement of experiments.

The modeling of biological systems needs to tackle the additional difficulty generated by the particular features of living matter. Among the various issues that will be critically analyzed in the next chapter, the lack of invariance principles that are typical of the inert matter systems. A critical analysis of this topic has been developed in the papers [Herrero (2007)], [May (2004)], and [Reed (2004)], see also [Hastings and Palmer (2003); Hastings *et al.* (2005)].

The paper by Hartwell *et al.* [Hartwell *et al.* (1999)] proposes a deep insight into the above issues and searches for a constructive interplay between the naive enthusiastic attitude of some applied mathematicians and the unreasonable scepticism of others. The main conceptual idea is that invariance principles are modified by the ability of living systems to express specific strategies that depend on survival purposes and adaptation to environmental conditions. Therefore, living systems have the ability to extract energy for their own well-being. Moreover, their adaptation ability generates mutations which occur at the molecular scale of genes and induce phenotype mutations. This evolution may even be very rapid in some specific pathologies.

It is worth stressing that the conclusive phrases of [Hartwell *et al.* (1999)] would seem to encourage a constructive interaction between biological and mathematical sciences. This concept is also related to the expectation of the scientific community which in this century, has been looking for a mathematical formalization of phenomena in life sciences that will be analogous to the mathematical formalization that characterized the progress of science over the last two centuries essentially devoted to the interaction between mathematical and physical sciences.

The above reasonings have motivated the contents of this monograph, whose main objective is to develop a mathematical theory of complex biological systems focusing on some specific systems, such as wound healing and repair, as well as some aspects of cancer phenomena. The concept of the mathematical theory is given in the next section. However, it is worth stating that we do not claim that this ambitious aim has been fully achieved; on the other hand, we hope that some preliminary results have been obtained and that this monograph offers guidelines for future developments.

1.2   On the Concept of Mathematical Theory

Let us first focus on the concept of the *mathematical theory of biological systems*, which should be clearly distinguished from that of *mathematical model*, whose derivation is based on conservation or equilibrium equations. Generally, these relations correspond to causality principles, closed by *phenomenological models* that describe the behavior of the matter. Mathematical models can be derived at any of the observation and representation scales that are typical of biological systems, namely molecular, cellular, and tissue scale.

Although tissues correspond to the largest macroscopic scale, the concept of *network* should be added as it technically refers to organs, where tissues are organized into complex systems that develop high-level specific functions that are connected through networks. The dynamics and the interplay among these networks is ruled at the lower cellular and molecular scales.

A mathematical theory should obtain the aforementioned material behaviors through a robust theory, delivered by biological sciences, that is suitable to transfer the information delivered at the molecular scale to the cellular scale. The mechanics of tissues is determined by the dynamics of cells, which are generated by the molecular dynamics. These concepts are reported in [Bellomo and Delitala (2008)] that focused on cancer phenomena, but they can easily be extended to a large variety of biological systems.

From the above reasoning, it is plain that the theory should specifically capture the multiscale aspects of all biological phenomena and select the correct mathematical framework to deal with the modeling at each scale. Subsequently, methods to link models at each scale have to be developed, for instance that of cells with those at the corresponding lower (genes) and higher (tissue) scales.

Considering that the existing mathematical methods have not yet been able to identify a uniquely defined approach, it is worth trying to understand why scientists are still far from developing a biological mathematical theory analogous to those of mathematical physics that have been developed over the last century. These concepts are specifically focused on the degeneration and repair of biological tissues, and, when appropriate, they are related to ageing phenomena.

1.3 Plan of the Monograph

After the previous preliminary introduction, the contents of this monograph, which is organized into nine more chapters, can be given.

- Chapter 2 deals with the analysis of the complexity characteristics of biological systems viewed as living systems. A reduction in complexity requires, as we shall see later, on the decomposition of the overall system into functional subsystems as the first step towards the development of the approach of systems biology. Subsequently, the approach derives evolution equations for each subsystem and the links among them to model the evolution of the overall system.

The other eight chapters are subdivided into three parts.

Part I is devoted to a phenomenological description of the class of biological systems under consideration in view of the mathematical approach proposed in the second and third parts of the monograph. A general strategy is identified, which is based on the concept that new mathematical tools should be developed to constrain the complexity features of living matter in equations, to the greatest possible extent.

- Chapter 3 presents a phenomenological description of the immune system, which plays an important role in the biological processes treated in the monograph (and, of course, in many other, if not all, processes). Specifically, this chapter is necessary for a deeper understanding of the subsequent one, which is devoted to the specific biological phenomena under consideration. The description is also inspired by the functional decomposition introduced in Chapter 2. Therefore, the components of the system are identified properly and analyzed focusing on the biological functions they express. The contents first introduce the role of bacteria and viruses, subsequently identify the main components of the immune systems, and finally analyze the respond of the immune, namely adaptive and acquired immunity. The presentation does not claim to be exhaustive. It simply aims at offering to applied mathematicians an overall presentation containing a reduced amount of information, which appears necessary to develop the modeling approach.

- Chapter 4 provides a phenomenological analysis of the wound healing process, in view of the development of the mathematical approach which needs a detailed focusing on well defined biological phenomena, so that it can be described by mathematical equations. The applications treated in Part III specifically refer to this chapter.

- Chapter 5 analyzes the different levels of organization related to the observation and representation scales from genes and mutations to wound healing, organ repair, and fibrosis diseases induced by cell mutations. The various concepts of the immune system introduced in Chapter 3 are here used for the aforementioned phenomenological description. This chapter also defines the general strategy that will be followed to pursue the objective of deriving a mathematical theory according to the approach of systems biology.

Part II provides a survey of the mathematical tools that will be used for the modeling applications. The contents takes advantage of the system biology approach and of the modeling strategy elaborated in Part I. The presentation focus on a revisiting and some developments of the existing literature in the field addressed to the specific aims of these Lectures Notes and to the applications that have been selected in Part III as case studies.

- Chapter 6 presents the derivation of the mathematical structures of the kinetic theory for active particles, known as the KTAP theory (for short), which is suitable to describe the dynamics of large systems of interacting living entities [Bellomo, Bianca, and Delitala (2009)]. This chapter also shows how the mathematical structures can be used towards the modeling at low scales, namely genes and cells which are regarded as independent systems. An introduction is above proposed from the applied mathematicians viewpoint, concerning the interpretation of the complex dynamics of immune competition. The analysis is developed at the cellular scale.

- Chapter 7 deals, at a methodological level, with the crucial problem of linking the dynamics from the small to the large scales, namely from genes to cells and from cells to tissues. Preliminarily, some phenomenological models of tissues are briefly reviewed to offer a panorama of the phenomenological models that will be compared with those derived from the underlying description at the cellular level. Subsequently, the mathematical asymptotic methods to derive macroscopic equations from the underlying equations at the microscopic scale, are presented. The method still needs some further developments of the tools presented in Chapter 6. The contents of this chapter complete the presentation of the mathematical tools and structures that can act as a general paradigm to derive specific models, and ultimately a biological mathematical theory.

Part III selects and develops some specific applications chosen according to the authors' bias. Moreover, it critically analyzes the central problem of deriving a mathematical theory of biological systems. Specifically, the strategy proposed in Part I is used to refer to the results obtained with the mathematical approach so that a variety of open problems are identified and brought to the attention of the reader.

- Chapter 8 deals with the application of the above mentioned approach to the analysis and the modeling of a specific biological phenomenon, namely the formation of a keloid triggered by viruses and the genetic susceptibility of a patient. The role of the immune system is also considered in this chapter, which also analyzes the possibility of the further degeneration of tissues related to inflammation followed by mutations that result in the onset of malignant phenomena.

- Chapter 9 shows how the methods mentioned in Chapter 7 can be developed properly and applied to derive models at the macroscopic scale, focusing on chemotaxis. This approach is an alternative to the classical continuum mechanics methods based on conservation equations closed by models of the material behavior of biological tissues. The application focuses on the derivation of chemotaxis models that play an important role in the wound healing processes.

- Chapter 10 proposes a critical analysis of the overall contents of the monograph. The various issues, developed in the preceding chapters, are viewed from a biological mathematical theory point of view, which is the main objective of this monograph. The critical analysis points out the conceptual difficulties which needs to be overcome towards the afore-said challenging research program.

The appendix offers some additional tools pertaining to the kinetic theory for active particles that are not specifically used in this monograph but which could be useful for further developments and applications. Several biological terms that should make applied mathematicians more acquainted with some topics of biological sciences are presented in a glossary. The glossary is not limited to the terms used in the monograph. Various terms has been added to improve the overall knowledge of the reader.

It is worth mentioning that the conceptual difficulty of the topics under consideration have motivated the design of new mathematical tools such as new concepts of system biology and further developments of the KTAP theory. The aim is twofold: to provide a new conceptual background and research perspectives for applied mathematicians involved in the challenging research field of mathematics for living systems, and to offer Lecture Notes for advanced courses in mathematical biology.

Chapter 2

On the Complexity of Biological Systems

2.1 Ten Common Features of Living Systems

This chapter deals with a phenomenological analysis of biological systems focusing on the particular features of living systems and their multiscale structure. Some general concepts of system biology are given and these are followed by an introduction on a general methodology to reduce complexity. The aim is to design some preliminary guidelines to help develops the mathematical approach that will be proposed in the following chapters.

All biological systems, and living systems in general, are constituted by several entities that interact in a nonlinear fashion. Understanding the role of nonlinear interactions is one of the greatest challenges in the study of complex systems considering that they are at the core of the emergence of qualitatively different states; namely, new states that are not mere combinations of those of the individual units comprising the system. In other words, the dynamics of each entity is not determined by the linear superposition of the action of all other entities, but their action acts strategically as a whole. This action depends on the strategy that they can develop, and differs with a variable number and localization of interacting entities.

In general, a system is an assembly of two or more elements, which interact to express a function as a whole. The parts of a system are interconnected and interdependent. Each system is composed of subsystems and it is nested within larger systems. The important issue to understand, whenever we talk about complex systems, is that everything and everyone are interconnected, and the whole has characteristics different from the parts.

Statistical physics can provide methods to study complex systems through techniques that are particularly suited for the study of systems with a large number of particles. However, the approach needs to be related to the fundamental concept of the quantitative characterization of system-scaling. On the other hand, the straightforward application of the methods of the physics of inert matter does not yet seem suitable to capture the essential features of the complexity of living systems. Moreover, it is necessary to have an appropriate insight into these methods to generate the development of a new mathematical approach, and possibly a mathematical theory.

Bearing all this in mind, let us list ten common complexity features, selected among several ones, that characterize living systems. The list does not claim to be exhaustive, but refers to the authors' personal experience and way of thinking.

1. A wide class of *complex system* is constituted by a large number of entities that will be later called *active particles* or *agents*. These elements are capable of interacting among themselves and with their outer environment. The overall system exhibits behaviors and characteristics that are different from those of the elements constituting the system. Ultimately, the *whole can be much more than the sum of its parts*.

2. *Interactions among active particles or agents* generally occur in a nonlinear fashion and involve immediate neighbors, but in some cases, also distant ones. Interactions can occur not only through contact, but also through distribution in space, considering that biological systems have the ability to communicate and may, in some cases, choose different observation paths. In other cases, topological distributions can play a remarkable role, for instance when active particles select a number of neighbors to communicate with, rather than selecting those, that are in their physical interaction domains.

3. *Heterogeneity* characterizes many living systems. In other words, the characteristics of active particles or agents can differ, even for entities with the same biological structure, due to, for instance, different phenotypes generated by the same genotype.

4. Interacting entities have the *ability to develop specific strategies*. This *self-organization ability* depends on the state of the surrounding environment, which is expressed even without the application of any external organizing principle. Generally, such an ability is heterogeneously distributed.

5. *Active particles or agents play a game at each interaction* with an output that is technically related to their strategy, which is often related to surviving and adaptation ability, and generally finalized to their own individual or collective well-being. In other words, interactions modify the state of the interacting entities according to the strategy they develop. Given a certain input, the output cannot, in general, be deterministically identified.

6. *Living systems usually operate out-of-equilibrium.* For example, living organisms are in a constant struggle with their environment to remain in a particular out-of-equilibrium state, namely alive. This dynamics is related, as already mentioned, to the search for well-being.

7. *Living systems* are generally constituted by a large number of *entities belonging to a broad variety of components*. Each component is constituted by entities that may occupy fixed positions or move freely in space or along networks. Therefore, these systems are different from those analyzed by statistical mechanics, which typically deals with systems containing many copies of a few interacting components. For instance, multicellular systems contain from millions to a few copies of each of thousands of different components, each with specific interactions.

8. *Time* plays an important role due to the *Darwinian type of cell mutations* and to the resolution process. The output appears to be caused by successive selections, which in some cases can be rapid, of entities, which become progressively resistent to a mutating environment. In some cases, mutations generate pathologies. *Time is a key variable* considering that the different rates that characterize biological processes and the ratio between them determines the output of the process.

9. The study of biological systems, in general of complex systems, always needs a *multiscale approach*. For instance, the dynamics of a cell at the molecular (genetic) level determines the cellular behaviors. Finally, the structures of the macroscopic tissues depends on such a dynamics. Ultimately, only one observation and representation scale is not sufficient to describe the dynamics of living systems.

10. *Small changes at the genetic or cellular scale can lead to large effects*, which in turn lead to ever larger effects. This snowballing dynamics is one thing that distinguishes living systems from mechanical systems, where small changes generally lead to small effects. This dynamics is also related to the fact that living systems receive feedback from their environments, which enables them to learn from their experiences.

The above list contributes to identify the conceptual difficulties that have to be tackled by a mathematical theory. The preliminary concept for applied mathematicians is the need of reducing complexity. This objective can be pursued by revisiting the basic concepts of systems biology, where, as already mentioned, the role of time has to be carefully considered to take into account the Darwinian selection related to the differentiation process schematically represented in Figure 2.1.

It is worth specifying that when we deal with biological systems it is also

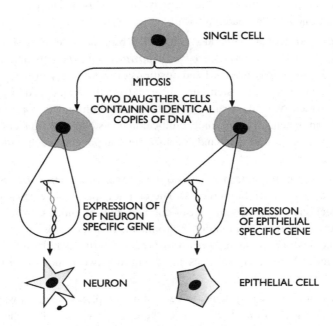

Fig. 2.1   A representation of the process of differentiation.

fundamental to consider the errors in the differentiation/replication that may generate cells that do not survive the contrast generated by the outer environment and/or the immune system. On the other hand, other cells may be better suited to survive and subsequently reproduce daughter cells with the same quality. Further mutations, called hallmarks [Hanahan and Weinberg (2000)], eventually generate cancer cells that may be cells which survive and proliferate.

2.2   Some Introductory Concepts of Systems Biology

This section provides a concise introduction to some preliminary concepts of systems biology aimed at offering a new interpretation of this approach, which will be developed in Chapter 3, based also on the various issues proposed in this section.

Systems biology is a new field of biological sciences that has the aim of developing a system-level understanding of biological phenomena [Kitano (2000); Woese (2004)]. This understanding requires a set of principles and methodologies that link the behaviors of molecules to system characteristics and functions. Ultimately, cells, organisms, and human beings can be described and understood at the system level on the basis of a consistent framework of knowledge that is underpinned by the basic principles of physics.

The scope of systems biology is potentially very broad and different types of techniques may be deployed for each research target. It requires collective efforts from multiple research areas, such as molecular biology, high-precision measurements, computer sciences, the control theory, and other scientific and engineering fields. Research needs to take advantage of methods developed in four key areas: *Genomic sciences* and molecular biology in general; Technology for *high-precision, comprehensive measurements*; Qualitative and quantitative analysis of the *dynamics of systems*, modeling each specific system under consideration; *computational sciences*, related to methods of bio-informatics, to obtain simulations in order to visualize phenomena and complete the qualitative analysis.

As already mentioned, a system is basically an assembly of components in a particular formation, yet it is more than a mere collection of components. To understand a system, it is necessary that its detailed description is followed by a deep comprehension of what happens when certain stimuli or disruptions occur.

The structures of systems need to be properly identified. Once the properties of a certain system have been identified up to a certain degree, its behavior needs to be understood. Various analysis methods can be used. For example, one may wish to know the sensitivity of certain behaviors against external perturbations, and how quickly the system returns to its normal state after the stimuli. Such an analysis not only reveals system-level characteristics, but also provides important insights for medical treatments by discovering the cell response to certain chemicals so that the effects can be maximized, while lowering possible side effects. In order to apply the insights obtained from the system structure and understand the behavior, research activity is needed to establish a method that can be used to control the dynamics of biological systems. Technologies to accomplish such control would enormously benefit human health.

There are various system structures that need to be identified, such as the structural relationship among cells in the developmental process, detailed cell-cell contact configuration, membrane, intra-cellular structures, and gene regulatory networks.

Simulations of the behavior of genes and metabolism networks play an important role in systems biology research. Due to the complexity of the network behavior and the fact that a large number of components are involved, it is almost impossible to understand intuitively the behaviors of such networks. In addition, accurate models and simulations are necessary prerequisite to analyze their dynamics by changing the parameters and structure of the gene and metabolism networks. Although such an analysis is necessary to understand the dynamics, these operations are not possible at the present state of the art in system biology.

Robustness is one of the essential features of systems biology. Understanding the mechanism behind robustness is particularly important because it provides in-depth understanding on how the system maintains its functional properties when faced with various disturbances. Specifically, we should be able to understand how organisms respond to: *changes in environment* (deprived nutrition level, *chemical attractant, exposure to various chemical agents that bind to receptors, temperature); Internal failures (DNA damage, genetic malfunctions in metabolic pathways)*.

A fundamental goal of systems biology consists in integrating different component parts of a biological system in order to understand the workings of the whole. Most biological systems exhibit a vast disparity of length and time scales or spatial scales and are inherently stochastic in nature.

The various genome and proteome projects [Human Genome (2005)], coupled with the advances and innovations made in microscopy and biological imaging, have provided descriptions, at the molecular and cellular levels, of unprecedented detail of the constituent parts and basic structures of living organisms. A full understanding of biological functions are achieved only if one is able to integrate all the relevant information at multiple of organization levels to recreate dynamic interactions. For example, Noble [Noble (2002)] explains how the success of drug therapy depends not only on the structure and direct function of the targeted protein but also on how it interacts with its surroundings: without this information it would not be possible to know which transporters, enzymes or receptors, are active in the diseased state and, furthermore, it would not be possible to predict (potentially fatal) side effects of the therapy that are due to these interactions. In general, these dynamic interactions cannot be recreated purely from experimental observations, and the only feasible approach is to develop mathematical and computational models which are able to couple together the underlying complex, nonlinear interacting processes.

The major challenge is to understand, at the system level, biological systems that are composed of components revealed through molecular biology.

2.3 Reducing Complexity

Let us focus on large aggregations of cells, namely multicellular systems, which are characterized by specific biological functions and show certain abilities, such as selective evolution adaptation to the environment, reproduction, competition and destruction and they can modify the functions they express.

The complexity sources of biological systems, which have been described in Section 2.1, indicate the need to reduce complexity. Several authors have suggested the approach of systems biology as an essential tool to achieve this objective. For instance, Woese [Woese (2004)] documents how system biology offers the key guidelines to deal with the interpretation and modeling of complex biological systems.

A useful way of reducing the complexity of the overall biological system is to decompose it into several interacting subsystems, each of them characterized by a lower order of complexity and characterized by well defined rules. Hartwell [Hartwell *et al.* (1999)] proposes the idea of reducing complexity by decomposing a biological system into several interacting modules, the so-called *theory of modules*. Hartwell's theory has been re-visited from the mathematical modeling point of view, in paper [Bellomo and Forni (2008)] which focuses on multicellular systems described by the kinetic theory for active particles [Bellomo (2008)]. In this paper, the authors suggest the term *functional subsystem* as an alternative to the term module. The component of a module, borrowing the terminology proposed in [Bellomo and Delitala (2008); Bellomo and Forni (2008)], is called an active particle.

Specifically, functional subsystems are identified by subsystems of cells that have the ability to collectively express a certain strategy, which is the same in each module and has to be identified by a scalar variable. The level of description of each module may not be the same for all of the modules. Moreover, interactions between different levels characterizes each element. The concept can be generalized to the lower molecular scale by grouping genes, whose expression collectively generates a certain phenotype.

Bearing all this in mind, the following statements can be made:

• A biological system is constituted by a large number of interacting entities, called *active particles*, whose microscopic state includes, in addition to geometrical and mechanical variables, an additional variable, called *activity*, which represents the individual ability to express a specific strategy.

• The activity variable is *heterogeneously* distributed over the active particles. Interactions modify the state of the interacting entities, while the strategy they express can be modified by the shape of their heterogeneous distribution.

• A *functional subsystem* is a collection of active particles that have the ability to express the same activity, which is regarded as a scalar variable. The whole system is constituted by several interacting functional subsystems.

• If the number of active particles in a functional subsystem is sufficiently large and the heterogeneity is a continuous variable, the state of the system can be described by a continuous *probability distribution*.

It is worth stressing that the link between a functional subsystem and its activity also depends on the specific phenomena that one aims at analyzing. Therefore, the decomposition into functional subsystems is a flexible approach that can be adapted to each particular investigation. Moreover, considering that the various subsystems are linked in networks, the modeling approach also needs to deal also with their interactions. The intensity of the expressed activity is generally different among the particles of the same functional subsystem.

The outlines given in this section, as already mentioned, should be regarded as preliminary concepts that can be transferred into mathematical equations, starting from Chapter 5. This mathematical approach can be regarded, as we shall see, as the natural development of methods of statistical mechanics so that active particles (rather than classical particles) have the ability to express specific strategies and interact in a nonlinear fashion.

PART 1

Immune System, Wound Healing Process, and System Biology

Chapter 3

The Immune System: A Phenomenological Overview

3.1 Introduction

The immune system is constituted of a network of cells, tissues, and organs that operate collectively to protect the body from bacterial, parasitic, fungal, viral infections, and from the growth of tumor cells. The cells involved are called white blood cells, or leukocytes, which act to seek out and destroy disease-causing organisms or substances. Many of these cell types have specialized functions and depend on the T helper subset for activation signals in the form of secretions formally known as cytokine, lymphokines, or more specifically interleukins. The cells of immune system attack, through a sequence of actions called the immune response, infectious agents and substances that invade the human body and cause diseases. The defence also applies to cells that, due to inflammation, may progressively degenerate into cancer cells (in general mutated cells).

Immune cells are able to perform complex tasks, such as *learning*, acquire the ability to distinguish between host entities (*self*) and foreign or infected entities (*non-self*), and *evolve* in time to progressively improve their overall defence ability, while retaining memory of previous encounters with foreign agents for a quicker response in the case of re-infection, and constantly updating their reactive potential.

The cells of the immune system, globally called *leukocytes* or *white blood cells*, communicate via cell-to-cell contact or via chemical signals by means of specific secreted substances, and cooperate continuously by monitoring the environment, detecting, and attacking foreign infectious agents.

The response of the immune system to an infectious agent can be subdivided into two main categories: *Innate immunity response*, which is characterized by the ability to efficiently discriminate between *self* and *non-self*, and *adaptive immune response* during which the immune system is able to learn and also

17

recognize cells that are carriers of a pathological state.

The question whether the immune system, which is specialized in detecting and eliminating foreign agents, is also able to recognize tumor cells, which are native to the body and substantially indistinguishable from normal cells, as *foreign*, has been a matter of discussions among biologists. Today, evidence is rapidly accumulating that the immune system contributes to the multilayered defenses of a body against tumors.

A phenomenological description of the immune system is here proposed, on the basis of that made by [Bellomo and Delitala (2008); Cooper (2010)] and to [Lollini, Motta, and Pappalardo (2006); Palladini (2010)], in view of the mathematical approach to modeling developed in the second part of this monograph. In particular, the modeling approach is focused on the following key issues: learning ability and progressive mutations.

The aim of this chapter is to review, in view of the modeling approach of a specific diseases, organs, cell types, interactions among cells, and deficiencies of the immune system, which is the body's defense against infectious organisms and other invaders. The contents are organized as follows: Section 3.2 provides a description of bacteria and viruses that can attack the immune system, whose components are described in Section 3.3. A phenomenological description of the immune response is delivered in Section 3.4 while the following section deals with a description of the diseases of the immune system and with a critical analysis on the overall contents.

3.2   Bacteria and Viruses

Bacteria are single-celled organisms that have no nucleus. They are 1/100th the size of a human cell and might measure 1 micrometer long. Bacteria are completely independent organisms able to eat and reproduce. Under the right conditions bacteria reproduce very quickly: One bacteria divides into two separate bacteria perhaps once every 20 or 30 minutes. At that rate, one bacteria can become millions in just a few hours.

A virus particle is a fragment of DNA (or RNA) in a protective coat. The virus comes in contact with a cell, attaches itself to the cell wall and injects its DNA (and perhaps a few enzymes) into the cell. The DNA uses the machinery inside the living cell to reproduce new virus particles. Eventually the hijacked cell dies and bursts, freeing the new virus particles; or the viral particles may bud off of the cell so it remains alive. In either case, the cell is a factory for the virus.

When a virus or bacteria (also known generically as a germ) invades the human body and reproduces, it normally causes problems. They cause things like colds, influenza, measles, mumps, malaria, AIDS and so on. For example, the strep throat bacteria (Streptococcus) releases a toxin that causes inflammation in the throat. The polio virus releases toxins that destroy nerve cells (often leading to paralysis). Some bacteria are benign or beneficial (for example, we all have millions of bacteria in our intestines and they help digest food), but many are harmful once they get into the body or the bloodstream.

Viral and bacterial infections are by far the most common causes of illness for most people. The job of the immune system is to protect the human body from these infections. It happens in three different ways:

• It creates a barrier that prevents bacteria and viruses from entering your body. Skin is an important part of the immune system. It acts as a primary boundary between germs and the human body. Skin is tough and generally impermeable to bacteria and viruses; the epidermis contains special cells called Langerhans cells (mixed in with the melanocytes in the basal layer) that are an important early-warning component in the immune system. The skin also secretes antibacterial substances.

• If a bacteria or virus does get into the body, the immune system tries to detect and eliminate it before it can make itself at home and reproduce. Tears and mucus contain an enzyme (lysozyme) that breaks down the cell wall of many bacteria. Saliva is also anti-bacterial. Since the nasal passage and lungs are coated in mucus, many germs not killed immediately are trapped in the mucus and soon swallowed. Mast cells also line the nasal passages, throat, lungs and skin. Any bacteria or virus that wants to gain entry to your body must first make it past these defenses.

• If the virus or bacteria is able to reproduce and start causing problems, the immune system is in charge of eliminating it.

3.3   The Immune System Components

The immune system is constituted of a network of cells, tissues, and organs that work together to protect the body. The major components of the immune system are: the lymphatic system, the white blood cells (leukocytes), the antibodies and the hormones. A detailed explanation of these components is given in the sequel.

3.3.1   *The Lymphatic System*

Lymph is an alkaline (pH > 7.0) fluid that is usually clear, transparent, and colorless. There are no red blood cells in lymph and it has a lower protein content than that of blood. It flows from the interstitial fluid through lymphatic vessels up to either the thoracic duct or right lymph duct (it drains the right sides of the thorax, neck, and head, whereas the thoracic duct drains the rest of the body), which terminate in the subclavian veins, where lymph is mixed into the blood. Lymph carries lipids and lipid-soluble vitamins absorbed from the gastrointestinal (GI) tract. Since there is no active pump in the lymph system, there is no back-pressure produced. The lymphatic vessels, like veins, have one-way valves that prevent backflow.

A target of the lymph system is to detect and remove bacteria. Small lymph vessels collect the liquid and move it toward larger vessels so that the fluid finally arrives to the lymph nodes that serve as filters of the lymphatic fluid. Antigen is usually presented to the immune system in the lymph nodes. When fighting certain bacterial infections, the lymph nodes swell with bacteria and the cells fighting the bacteria, to the point where you can actually feel them. Swollen lymph nodes are therefore a good indication that a person has an infection of some sort. Once lymph has been filtered through the lymph nodes it re-enters the bloodstream.

The human lymphoid system is constituted by the following organs:

• bone marrow (in the hollow center of bones) and the thymus gland (located behind the breastbone above the heart) called primary organs;

• adenoids, tonsils, spleen (located at the upper left of the abdomen), lymph nodes (along the lymphatic vessels with concentrations in the neck, armpits, abdomen, and groin), Peyer's patches (within the intestines), and the appendix. These organs are called secondary organs.

All the cells of the immune system are initially derived from the bone marrow. They form through a process called hematopoiesis, during which bone marrow-derived stem cells (they are called "stem cells" because they can branch off and become many different types of cells) differentiate into either mature cells of the immune system or into precursors of cells that migrate out of the bone marrow to continue their maturation elsewhere. The bone marrow produces specific immune cells (B cells, natural killer cells, granulocytes and immature thymocytes, see the next subsection for their explanation), in addition to red blood cells and platelets.

The thymus lives in the chest, between the breast bone and the heart. It is responsible for producing another type of immune cells (the T-cells), and is especially important in newborn babies - without a thymus a baby's immune system collapses and the baby will die. The thymus seems to be much less important in adults - for example, you can remove it and an adult will live because other parts of the immune system can handle the load. However, the thymus is important, especially to T cell maturation (a remarkable maturation process sometimes referred to as thymic education). The mature T cells are then released into the bloodstream.

The spleen is an immunologic filter of the blood by foreign cells. It is made up of B cells, T cells, macrophages, dendritic cells, natural killer cells and red blood cells. In addition to capturing foreign materials (antigens) from the blood that passes through the spleen, migratory macrophages and dendritic cells bring antigens to the spleen via the bloodstream. This organ can be thought of as an immunological conference center.

3.3.2 The White Blood Cells

The white blood cells (leukocytes) are probably the most important part of the immune system and constitute an overall collection of different cells that work together to destroy pathogenic agents. Leukocytes actually act like independent, living single-cell organisms able to move and engulf viruses and bacteria. They behave very much like amoeba in their movements and many of them cannot divide and reproduce on their own, instead have a factory somewhere in the body that produces them: the bone marrow.

Indeed, these cells start in bone marrow as stem cells. As already mentioned, stem cells are generic cells that can form into the many different types of leukocytes as they mature. The stem cells will divide and differentiate into all different types of white blood cells. A "bone marrow transplant" is accomplished simply by injecting stem cells from a donor into the blood stream, they find their way, almost magically, into the marrow and make their home there.

Leukocytes are divided into two main classes:

1) **Granulocytes**, which get their name because they contain granules in the cytoplasm, make up 50% to 60% of all leukocytes. Granulocytes are composed of three cell types identified as *neutrophils*, *eosinophils*, and *basophils*. These cells are predominantly important in the removal of bacteria and parasites from the body. They engulf these foreign bodies and degrade them using their powerful enzymes.

2) **Agranulocytes**, which are without granules. This class include *lymphocytes* and *monocytes*. Lymphocytes that make up 30% to 40% of all leukocytes, come in two classes: B cells (those that mature in bone marrow) and T cells (those that mature in the thymus) which are divided in *helper*, *killer*, and *suppressor*. The lymphocytes that handle most of the bacterial and viral infections that we get, start in the bone marrow. T cells and B cells are often found in the bloodstream

but tend to concentrate in lymph tissue such as the lymph nodes, the thymus, and the spleen. There is also quite a bit of lymph tissue in the digestive system. Monocytes that make up 7% or so of all leukocytes, evolve into *macrophages*.

Each of the different types of white blood cells have a special role in the immune system, and many are able to transform themselves in different ways. The following descriptions help to understand the roles of the different cells.

• Polymorphonuclear neutrophils, also called polys for short, are by far the most common form of white blood cells of the human body. The bone marrow produces trillions of them every day and releases them into the bloodstream, but their life span is short (half-life of 6-8 hours, 1-4 day lifespan). Poly are phagocytes that have no mitochondria and get their energy from stored glycogen. Moreover they are nondividing and have a segmented nucleus. The neutrophils provide the major defense against pyogenic (pus-forming) bacteria and are the first on the scene to fight infection. Once a neutrophil finds a foreign particle or a bacteria it will engulf it, releasing enzymes, hydrogen peroxide and other chemicals from its granules to kill the bacteria. In a site of serious infection (where lots of bacteria have reproduced in the area), pus will form. Pus is simply dead neutrophils and other cellular debris. Once in the bloodstream neutrophils can move through capillary walls into tissue. If you get a splinter or a cut, neutrophils will be attracted by a process called chemotaxis. Poly are followed by the wandering macrophages about three to four hours later.

• Eosinophils and basophils are far less common than neutrophils. Eosinophils seem focused on parasites in the skin and the lungs, while basophils carry histamine and therefore important (along with mast cells) to causing inflammation. From the immune system's standpoint inflammation is a good thing. It brings in more blood and it dilates capillary walls so that more immune system cells can get to the site of infection.

• B cells are the cells that produce antibodies in response to foreign proteins of bacteria, viruses, and tumor cells. When stimulated (activated in the spleen), B cells mature into plasma cells and produce large amounts of antibody. A specific B cell is tuned to a specific germ, and when the germ is present in the body, the B cell clones itself and produces millions of antibodies designed to eliminate the germ.

• T cells, on the other hand, actually bump up against cells and kill them. They are usually divided into three major subsets that are functionally and phenotypically (identifiably) different: Killer (or cytotoxic, CD8+), helper (CD4+), and suppressor. Here (CD) means cluster of differentiation.
Killer T cells can detect cells that are harboring viruses, and when it detects such

a cell it kills it. These cells are important in directly killing certain tumor cells, viral-infected cells and sometimes parasites.

Helper and Suppressor T cells help sensitize killer T cells and control the immune response. Suppressor T cells are also important in down-regulation of immune responses. The main function of the T helper cell is to augment or potentiate immune responses by the secretion of specialized factors that activate other white blood cells to fight off infection. Helper and Suppressor T cells often depend on the secondary lymphoid organs (the lymph nodes and spleen) as sites where activation occurs, but they are also found in other tissues of the body, most conspicuously the liver, lung, blood, and intestinal and reproductive tracts.

Helper T cells are actually quite important and interesting. They are activated by Interleukin-1, produced by macrophages. Once activated, Helper T cells produce Interleukin-2, then interferon and other chemicals. These chemicals activate B cells so that they produce antibodies.

• **Natural killer cells**, often referred to as NK cells, are similar to the killer T cell subset. They function as effector cells that directly kill certain tumors such as melanomas, lymphomas and viral-infected cells, most notably herpes and cytomegalovirus-infected cells. NK cells, unlike the CD8+ (killer) T cells, kill their targets without a prior "conference" in the lymphoid organs. However, NK cells that have been activated by secretions from CD4+ T cells will kill their tumor or viral-infected targets more effectively.

• **Macrophages** are the biggest (hence the name "macro") white blood cells and are important in the regulation of immune responses. Monocytes are released by the bone marrow, float in the bloodstream, enter tissue and turn into macrophages. Most boundary tissue has its own devoted macrophages. For example, alveolar macrophages live in the lungs and keep the lungs clean (by ingesting foreign particles like smoke and dust) and disease free (by ingesting bacteria and microbes). Macrophages are called langerhans cells when they live in the skin. One of their jobs is to clean up dead neutrophils – macropghages clean up pus, for example, as part of the healing process. Macrophages are often referred to as scavengers or antigen-presenting cells (APC) because they pick up and ingest foreign materials and present these antigens to other cells of the immune system such as T cells and B cells. This is one of the important first steps in the initiation of an immune response. Stimulated macrophages exhibit increased levels of phagocytosis and are also secretory. The complexity and level of interaction between neutrophils, macrophages, T cells and B cells is really quite amazing.

• **Dendritic cells** (addressed only recently), which also originate in the bone marrow, function as antigen presenting cells (APC). In fact, the dendritic cells are more efficient APCs than macrophages. These cells are usually found in the structural compartment of the lymphoid organs such as the thymus, lymph nodes, and spleen. However, they are also found in the bloodstream and other tissues of the body. It is believed that they capture antigen or bring it to the lymphoid organs where an immune response is initiated. Unfortunately, dendritic cells are extremely hard to isolate, which is often a prerequisite for the study of the functional qualities of specific cell types. Of particular issue here is the recent finding that dendritic cells bind high amount of HIV, and may be a reservoir of virus that is transmitted to CD4+ T cells during an activation event.

3.3.3 *Antibodies and Hormones*

Antibodies are specialized proteins that specifically recognize and bind to one particular protein. Antibody production and binding to a foreign substance or antigen, often is critical as a means of signaling other cells to engulf, kill or remove that substance from the body.

Antibodies (also referred to as immunoglobulins and gammaglobulins) are produced by white blood cells. They are Y-shaped proteins that each respond to a specific antigen (bacteria, virus or toxin). Each antibody has a special section (at the tips of the two branches of the Y) that is sensitive to a specific antigen and binds to it in some way. When an antibody binds to a toxin, it is called an antitoxin (if the toxin comes from some form of venom, it is called an antivenin). The binding generally disables the chemical action of the toxin. When an antibody binds to the outer coat of a virus particle or the cell wall of a bacterium it can stop their movement through cell walls. Or a large number of antibodies can bind to an invader and signal to the complement system that the invader needs to be removed. There are five class of antibodies: Immunoglobulin A (IgA), immunoglobulin D (IgD), immunoglobulin E (IgE), immunoglobulin G (IgG), immunoglobulin M (IgM).

There are several hormones generated by components of the immune system. These hormones are known generally as lymphokines. It is also known that certain hormones in the body suppress the immune system, e.g. steroids and corticosteroids (components of adrenaline). Tymosin (thought to be produced by the thymus) is a hormone that encourages lymphocyte production. Interleukins are another type of hormone generated by white blood cells. For example, Interleukin-1 is produced by macrophages after they eat a foreign cell. IL-1 has an interesting side-effect - when it reaches the hypothalamus it produces fever and fatigue.

Tumor Necrosis Factor (TNF) is a hormone produced by macrophages. It is able to kill tumor cells, and it also promotes the creation of new blood vessels so it is important to healing.

Interferon interferes with viruses (hence the name) and is produced by most cells in the body. Interferons, like antibodies and complements, are proteins, and their job is to let cells signal to one another. When a cell detects interferon from other cells, it produces proteins that help prevent viral replication in the cell.

3.4   The Immune Response

Important questions to ask about white blood cells (and several other parts of the immune system) are:

How does a white blood cell know what to attack and what to leave alone?

Why doesn't a white blood cell attack every cell in the body?

There is a system built into all of the cells in the human body called the *Major Histocompatibility Complex* (MHC), also known as the *Human Leukocyte Antigen* (HLA), that marks the cells in your body as "self". Anything that the immune system finds that does not have these markings (or that has the wrong markings) is definitely "non self" and is therefore fair game. There are two major types of MHC protein molecules–class I and class II–that span the membrane of almost every cell in an organism. In humans these molecules are encoded by several genes all clustered in the same region on chromosome 6. Each gene has an unusual number of alleles (alternate forms of a gene). As a result, it is very rare for two individuals to have the same set of MHC molecules, which are collectively called a tissue type.

The response of the immune system to an infectious agent is subdivided into two main categories:

- *Innate (non-specific) immunity response*, which is mediated by granulocytes, macrophages, and NK cells;

- *Adaptive (specific, acquired) immune response*, which is mediated by the *lymphocytes*.

Although the innate and adaptive immune responses both operate to protect against invading organisms, they differ in a number of ways. The innate immune system is constitutively active and reacts immediately to infection. The adaptive immune response to an invading organism takes some time to develop. The innate immune system is not specific in its response and reacts equally well to a variety of organisms, whereas the adaptive immune system is antigen-specific and reacts only with the organism that induced the response. Finally, the adaptive immune

Table 3.1   The fundamental differences between the innate and adaptive immune system.

INNATE IMMUNITY	ADAPTIVE IMMUNITY
Response is antigen-independent	Response is antigen-dependent
There is immediate maximal response	There is a lag time between exposure and maximal response
Not antigen-specific	Antigen-specific
Exposure results in no immunologic memory	Exposure results in immunologic memory

system exhibits immunological memory. It "remembers" that it has encountered an invading organism (antigen) and reacts more rapidly on subsequent exposure to the same organism. The innate immune system does not possess a memory. Table 3.1 summarizes the fundamental differences between the innate and adaptive immune response.

It is worth mentioning the concept of *repertoire* of an immune system, which is a set of antigen receptors each having a unique specificity to bind an antigen. In many vertebrate species, antigen receptors are produced via combinatorial arrangements of DNA segments in specialized immune cells. Due to this molecular mechanism, repertoire of vertebrate species is potentially very large. The diversity of repertoire is thought to guarantee recognition of most ill-causing microorganisms, see [Anderton and Wraith (2002); Schwartz (2003)].

The above aspects need to be carefully considered in the modeling approach. In fact, the evolution of pathological states depends on the speed by which immune cells recognize cells that are carriers of a pathological state. A detailed analysis of the difference between the above two types of immunity is given in the already cited paper [Cooper (2010)], while a brief, however exhaustive description of the dynamics of the immune system at the cellular scale is given in [Lollini, Motta, and Pappalardo (2006)].

3.4.1   *Innate Immunity*

The elements of the innate (non-specific) immune system include anatomical barriers, secretory molecules and cellular components. Among the mechanical anatomical barriers are the skin and internal epithelial layers, the movement of the intestines and the oscillation of broncho-pulmonary cilia. Associated with these protective surfaces are chemical and biological agents.

Anatomical barriers to infections. The epithelial surfaces form a physical barrier that is very impermeable to most infectious agents. Thus, the skin acts as our first line of defense against invading organisms. The desquamation of skin epithelium also helps remove bacteria and other infectious agents that have adhered to the epithelial surfaces. Movement due to cilia or peristalsis helps to keep air passages and the gastrointestinal tract free from microorganisms. The flushing action of tears and saliva helps prevent infection of the eyes and mouth. The trapping effect of mucus that lines the respiratory and gastrointestinal tract helps protect the lungs and digestive systems from infection.

Fatty acids in sweat inhibit the growth of bacteria. Lysozyme and phospholipase found in tears, saliva and nasal secretions can breakdown the cell wall of bacteria and destabilize bacterial membranes. The low pH of sweat and gastric secretions prevents growth of bacteria. Defensins (low molecular weight proteins) found in the lung and gastrointestinal tract have antimicrobial activity. Surfactants in the lung act as opsonins (substances that promote phagocytosis of particles by phagocytic cells).

The normal flora of the skin and in the gastrointestinal tract can prevent the colonization of pathogenic bacteria by secreting toxic substances or by competing with pathogenic bacteria for nutrients or attachment to cell surfaces.

Humoral barriers to infection. The anatomical barriers are very effective in preventing colonization of tissues by microorganisms. However, when there is damage to tissues, the anatomical barriers are breached and infection may occur. Once infectious agents have penetrated into tissues, another innate defense mechanism comes into play, namely acute inflammation. Humoral factors play an important role in inflammation, which is characterized by edema and the recruitment of phagocytic cells. The following humoral factors are found in serum or they are formed at the site of infection:

• The *complement system* is the major humoral non-specific defense mechanism (see complement chapter). Once activated complement can lead to increased vascular permeability, recruitment of phagocytic cells, and lysis and opsonization of bacteria.

• Depending on the severity of the tissue injury, the *coagulation system* may or may not be activated. Some products of the coagulation system can contribute to the non-specific defenses because of their ability to increase vascular permeability and act as chemotactic agents for phagocytic cells. In addition, some of the products of the coagulation system are directly antimicrobial. For example, beta-lysin, a protein produced by platelets during coagulation can lyse many Gram positive bacteria by acting as a cationic detergent.

• *Lactoferrin* and transferrin. By binding iron, an essential nutrient for bacteria, these proteins limit bacterial growth.

• *Interferons* are proteins that can limit virus replication in cells.

•. *Lysozyme* down the cell wall of bacteria.

• *Interleukin-1* induces fever and the production of acute phase proteins, some of which are antimicrobial because they can opsonize bacteria.

Cellular barriers to infection. Each of the cells in the innate immune system bind to antigen using pattern-recognition receptors. These receptors are encoded in the germ line of each person. This immunity is passed from generation to generation. Over the course of human development these receptors for pathogen-associated molecular patterns have evolved via natural selection to be specific to certain characteristics of broad classes of infectious organisms. There are several hundred of these receptors and they recognize patterns of bacterial lipopolysaccharide, peptidoglycan, bacterial DNA, dsRNA, and other substances.

Part of the inflammatory response is the recruitment of polymorphonuclear eosinophiles and macrophages to sites of infection. These cells are the main line of defense in the innate immune system.

Neutrophils are recruited to the site of infection where they phagocytose invading organisms and kill them intracellularly. In addition, they contribute to collateral tissue damage that occurs during inflammation.

Tissue macrophages and newly recruited monocytes, which differentiate into macrophages, also function in phagocytosis and intracellular killing of microorganisms. In addition, macrophages are capable of extracellular killing of infected or altered self target cells. Furthermore, macrophages contribute to tissue repair and act as antigen-presenting cells, which are required for the induction of specific immune responses.

Natural killer (NK) and lymphokine activated killer (LAK) cannot specifically kill virus infected and tumor cells. These cells are not part of the inflammatory response but they are important in adaptive immunity to viral infections and tumor surveillance.

Eosinophils have proteins in granules that are effective in killing certain parasites.

It is worth to pointing out that the characteristics of the innate immune response include: Responses are broad-spectrum (non-specific), there is no memory or lasting protective immunity, there is a limited repertoire of recognition molecules, and the responses are phylogenetically ancient.

3.4.2 Adaptive Immunity

The adaptive specific immune response acts only if these innate defenses are breached. Adaptive (acquired) immunity refers to antigen-specific defense mechanisms that take several days to become protective and are designed to remove a specific antigen. There are two major branches of the adaptive immune responses:

-*Humoral immunity*, which involves the production of antibody molecules in response to an antigen and is mediated by B-lymphocytes;

- *Cell-mediated immunity*, which involves the production of killer T cells, activated macrophages, activated NK cells, and cytokines in response to an antigen.

Although adaptive immunity develops in an animal which is undergoing a specific immunological response to an antigen, the immune cells and factors generated can be shared among two or more animals. Hence, adaptive immunity can be acquired by an animal in two ways:

-*Active immunity*. The animal undergoes an immunological response to an antigen and produces the cells and factors responsible for the immunity, i.e., the animal produces its own antibodies and/or immuno-reactive lymphocytes. Active immunity can persist a long time in the animal, up to many years in humans.

-*Passive immunity* is the acquisition by an animal of immune factors which were produced in another animal, i.e., the host receives antibodies and/or immuno-reactive lymphocytes originally produced during an active response in another animal. Passive immunity is typically short-lived and usually persists for only a few weeks or months.

It is worth pointing out that either active or passive immunity may be acquired by natural means (e.g. self production of antibodies during infection or transfer of antibodies from mother to offspring) or by artificial means (i.e., vaccination and other immunization procedures).

Let us describe the process of recognition of the adaptive immune response (see Figure 3.1 for a synthetic description).

The first step is done by Antigen-Presenting Cells (APCs), dendritic cells or macrophages, which capture protein antigens through pinocytosis and phagocytosis. Upon capturing antigens and becoming activated by proinflammatory cytokines, the dendritic cells detach from the epithelium, enter lymph vessels, and are carried to regional lymph nodes. By the time they enter the lymph nodes, they have matured and are now able to present antigen to the everchanging populations

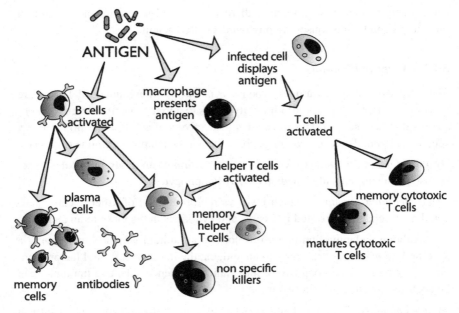

Fig. 3.1   The recognition process of the adaptive immune system: Under antigenic stimulus, B-lymphocytes interact with macrophages and T helper cells, and proliferate, differentiate into antibody-secreting plasma cells. The plasma cells synthesize large amounts of immunoglobulins (antibodies) which will react stereochemically with the stimulating antigen. The secreted antibody binds to the antigen and in some way leads to its neutralization or elimination from the body. B-lymphocytes and T-lymphocytes also develop into clones of identical reactive cells called memory B-cells and T-cells.

of naive T-lymphocytes (those have not yet encountered an antigen) located in the cortex of the lymph nodes.

Helper T-lymphocytes have receptors that bind molecules made primarily by antigen-presenting cells (MHC-II molecules). Each helper T-lymphocyte is genetically programmed to make a unique recepetor.

Most proteins are T-dependent antigens. In order to have naive B-lymphocytes (those have not yet encountered an antigen) proliferate, differentiate, and mount an antibody response against T-dependent antigens, these B-lymphocytes must interact with the helper T-lymphocytes. All classes of antibody molecules can be made against T-dependent antigens and there is usually a memory response against such antigens. A few antigens are called T-independent antigens. T-independent antigens are usually large carbohydrate and lipid molecules with multiple, repeating subunits. B-lymphocytes mount an antibody response to T-independent antigens without the requirement of interaction with T helper cells.

The resulting antibody molecules do not give rise to a memory response.

Under antigenic stimulus, B-lymphocytes become transformed into antibody-secreting plasma cells. The plasma cells synthesize large amounts of immunoglobulins (antibodies) which will react stereochemically with the stimulating antigen.

After the naive B-lymphocytes, helper T-lymphocytes, and Killer T-lymphocytes bind their corresponding epitopes, they must proliferate into large clones of identical cells in order to mount a successful immune response against that antigen. This is known as clonal expansion. In the vast majority of immune responses, the B-lymphocytes and T-lymphocytes that were activated must be stimulated to proliferate by means of cytokines (such as IL-2, IL-4, IL-5, Il-6, and IL-10) produced primarily by helper T-lymphocytes.

It is thought that in most immune responses, only around 1/1000 to 1/10,000 lymphocytes will have a receptor capable of binding the initiating antigen. Thus, proliferation allows the production of clones of thousands of identical lymphocytes having specificity for the original antigen. This is essential to give enough cells to mount a successful immune response against that antigen.

Antibodies, cytokines, activated macrophages, and T killer-lymphocytes eventually destroy or remove the antigen. Antibodies and cytokines amplify defense functions and collaborate with cells of the innate immune system, such as phagocytes and NK cells, as well as with molecules of the innate immune system, such as those of the complement system and the acute phase response. T killer-lymphocytes destroy body cells having the original epitope on their surface, e.g., viral infected cells, bacterial infected cells, and tumor cells. Cytokines also amplify innate immune defenses such as inflammation, fever, and the acute phase response.

During the proliferation and differentiation that follows lymphocyte activation, some of the B-lymphocytes and T-lymphocytes do not generate antibodies, clone themselves, and become circulating, long-lived memory cells. Memory cells are capable of what is called anamnestic response or "memory", that is, they "remember" the original antigen. These cells remains in the human body for years, so if that same antigen again enters the body while the memory cells are still present, these memory cells will initiate a rapid, heightened secondary response against that antigen.

This is why the body sometimes develops a permanent immunity after an infectious disease and is also the principle behind immunization.

The immune responses are carefully regulated by a variety of mechanisms. They are turned on only in response to an antigen and are turned off once the antigen has been removed. Suppressor T-lymphocytes play an important role in turning on and turning off immune responses. If there is a breakdown in this normal elimination or suppression of self-reacting cells, autoimmune diseases may develop.

We conclude this description summarizing, in view of the next section, the three important features of an immunological response relevant to host defense and/or "immunity" to pathogenic microorganisms:

• *Specificity.* An antibody or reactive T-lymphocyte reacts specifically with the antigen that induced its formation; it does not react with other antigens. Generally, this specificity is of the same order as that of enzyme-substrate specificity or a receptor-ligand interaction. However, cross-reactivity is possible. The specificity of the immunological response is explained on the basis of the clonal selection hypothesis: during the primary immune response, a specific antigen selects a pre-existing clone of specific lymphocytes and stimulates exclusively its activation, proliferation and differentiation.

• *Memory.* The immunological system has a "memory". Once the immunological response has reacted to produce a specific type of antibody or reactive T-cell, it is capable of producing more of the antibody or activated T-lymphocytes rapidly and in larger amounts. This is referred to as a secondary or memory response.

• *Tolerance.* An animal generally does not undergo an immunological response to its own (potentially-antigenic) components. The animal is said to be tolerant to self-antigens. This ensures that under normal conditions, an immune response to "self" antigens (called an autoimmune response) does not occur. Autoimmune responses are potentially harmful to the host. Tolerance is brought about in a number of ways, but basically the immunological system is able to distinguish "self" antigens from "non-self" (foreign) antigens; it will respond to "non-self" but not to "self". Sometimes in an animal, tolerance can be "broken", which may result in an autoimmune pathology.

3.5   Immune System Diseases

Occasionally the immune system makes a mistake. Disorders of the immune system fall into four main categories:

(1) Immunodeficiency disorders (primary or acquired);
(2) Autoimmune disorders, in which the body's own immune system attacks its own tissue as foreign matter);

(3) Allergic disorders (in which the immune system overreacts in response to an antigen);

(4) Cancers of the immune system.

Immunodeficiencies occur when a part of the immune system is not present or is not working properly. Sometimes a person is born with an immunodeficiency (known as primary immunodeficiencies), although symptoms of the disorder might not appear until later in life. Immunodeficiencies also can be acquired through infection or produced by drugs (these are sometimes called secondary immunodeficiencies). Immunodeficiencies can affect B lymphocytes, T lymphocytes, or phagocytes.

Examples of primary immunodeficiencies are:

• IgA deficiency is the most common immunodeficiency disorder. IgA is an immunoglobulin that is found primarily in the saliva and other body fluids that help guard the entrances to the body. IgA deficiency is a disorder in which the body does not produce enough of the antibody IgA. People with IgA deficiency tend to have allergies or get more colds and other respiratory infections, but the condition is usually not severe.

• Severe combined immunodeficiency (SCID) is also known as the "bubble boy disease" after a Texas boy with SCID who lived in a germ-free plastic bubble. SCID is a serious immune system disorder that occurs because of a lack of both B and T lymphocytes, which makes it almost impossible to fight infections.

• DiGeorge syndrome (thymic dysplasia), a birth defect in which kids are born without a thymus gland, is an example of a primary T-lymphocyte disease. The thymus gland is where T lymphocytes normally mature.

• Chediak-Higashi syndrome and chronic granulomatous disease both involve the inability of the neutrophils to function normally as phagocytes.

Acquired (or secondary) immunodeficiencies usually develop after someone has a disease, although they can also be the result of malnutrition, burns, or other medical problems. Acquired (secondary) immunodeficiencies include:

• HIV (human immunodeficiency virus) infection/AIDS (acquired immunodeficiency syndrome). It is a disease that slowly and steadily destroys the immune system. It is caused by HIV, a virus that wipes out the T-helper cells. Without T-helper cells, the immune system is unable to defend the body against normally harmless organisms, which can cause life-threatening infections in people who have AIDS. Newborns can get HIV infection from their mothers while in the uterus, during the birth process, or during breastfeeding. People can get

HIV infection by having unprotected sexual intercourse with an infected person or from sharing contaminated needles for drugs, steroids, or tattoos.

• **Immunodeficiencies caused by medications.** Some medicines suppress the immune system. One of the drawbacks of chemotherapy treatment for cancer, for example, is that it not only attacks cancer cells, but also other fast-growing, healthy cells, including those found in the bone marrow and other parts of the immune system. In addition, people with autoimmune disorders or who have had organ transplants may need to take immunosuppressant medications, which can also reduce the immune system's ability to fight infections and can cause secondary immunodeficiency.

Autoimmune disorders. The immune system mistakenly attacks the body's healthy organs and tissues as though they were foreign invaders. Autoimmune diseases include:

• **Lupus,** a chronic disease marked by muscle and joint pain and inflammation (the abnormal immune response also may involve attacks on the kidneys and other organs)

• **Juvenile rheumatoid arthritis,** a disease in which the body's immune system acts as though certain body parts (such as the joints of the knee, hand, and foot) are foreign tissue and attacks them

• **Scleroderma,** a chronic autoimmune disease that can lead to inflammation and damage of the skin, joints, and internal organs

• **Ankylosing spondylitis,** a disease that involves inflammation of the spine and joints, causing stiffness and pain

• **Juvenile dermatomyositis,** a disorder marked by inflammation and damage of the skin and muscles

Allergic disorders occur when the immune system overreacts to exposure to antigens in the environment. The substances that provoke such attacks are called allergens. The immune response can cause symptoms such as swelling, watery eyes, and sneezing, and even a life-threatening reaction called anaphylaxis. Medications called antihistamines can relieve most symptoms. Allergic disorders include:

• **Asthma,** a respiratory disorder that can cause breathing problems, frequently involves an allergic response by the lungs. If the lungs are oversensitive to certain allergens (like pollen, molds, animal dander, or dust mites), it can trigger breathing tubes in the lungs to become narrowed, leading to reduced airflow and making it hard for a person to breathe.

- **Eczema** is an itchy rash also known as atopic dermatitis. Although atopic dermatitis is not necessarily caused by an allergic reaction, it more often occurs in kids and teens who have allergies, hay fever, or asthma or who have a family history of these conditions.

- **Allergies of several types can occur in kids and teens.** Environmental allergies (to dust mites, for example), seasonal allergies (such as hay fever), drug allergies (reactions to specific medications or drugs), food allergies (such as to nuts), and allergies to toxins (bee stings, for example) are the common conditions people usually refer to as allergies.

Cancer occurs when cells grow out of control. This can also happen with the cells of the immune system. Lymphoma involves the lymphoid tissues and is one of the more common childhood cancers. Leukemia, which involves abnormal overgrowth of leukocytes, is the most common childhood cancer. With current medications most cases of both types of cancer in kids and teens are curable.

It is worth mentioning that another example of an immune system mistake is the effect the immune system has on transplanted tissue. This really isn't a mistake, but it makes organ and tissue transplants nearly impossible. When the foreign tissue is placed inside your body, its cells do not contain the correct identification. The immune system therefore attacks the tissue. The problem cannot be prevented, but can be diminished by carefully matching the tissue donor with the recipient and by using immuno-suppressing drugs to try to prevent an immune system reaction. Of course, by suppressing the immune system these drugs open the patient to opportunistic infections.

3.6 Critical Analysis

The immune system is considered one of the most significant existing representations of complexity. In fact, we have seen that a large variety of external agents, for instance bacteria and viruses, have to be considered before developing an insight into the several components of the system under consideration. Moreover, the system itself can degenerate, for instance due to mutations, which generate new components that contrast further the system. All these components express well defined biological functions with heterogeneously distributed characteristics due to different type of mutations and of the time when are generated.

However limited the description delivered in this chapter may appear from the viewpoint of biology, it already farsees the difficulty in transferring such description into mathematical equations. In fact, mathematicians should consider a large number of equations corresponding to the number of components, complex interactions, the randomness induced by heterogeneity, the onset of new components, hence of new equations due to mutations.

All the above reasonings suggest that the traditional approach of system biology needs to be further developed to take into account all the above aspects with the final aim of reducing complexity without, however, losing the essence of biological characteristics that are expressed at the molecular and cellular scale. Therefore this chapter offers, as the next one, some reasonings on biological systems that can contribute to the above outlined development of new concepts of system biology.

As we have seen the preceding descriptions have been focused on specific biological phenomena and systems. However, the presence of the immune system needs to be considered for any type of phenomena we consider. In fact, all pathology or undesired presence, activate the defence action of the body.

Accordingly, the mathematical model of keloid formation, triggered by a virus, and degeneration includes the modeling of the immune action by an approach where the overall complexity has been greatly reduced according to the guidelines delivered in Chapter 5. We anticipate, in view of further reasonings, that the modeling approach can ever be refined by inserting the mathematical description of additional specific phenomena.

Chapter 4

Wound Healing Process and Organ Repair

4.1 Introduction

The demand for tissue and organ replacement as a consequence of tissue damage or diseases is rapidly growing. Moreover, while the number of patients suffering from organ failure is increasing, there is a significant shortage of organs available for transplantation. The partial lack of organ functionality and non-severe tissue damage, although not immediately life-threatening and not requiring a transplant, also has important consequences on the quality of life of patients, and constitutes a source of high costs over a prolonged period. Although less famous than cancer, the problem of abnormal organ repair is however also important, as far as health care is concerned.

This chapter offers a phenomenological description of the wound healing, fibrosis, and recovery processes. Genetic mutations can occur during this process and trigger the onset of fibrosis diseases. The description does not claim to be exhaustive. Surely, it is not satisfactory from a biology point of view, however a collection of brief descriptions is organized in view of the modeling approach that should depict the time evolution of the various phenomena under consideration. In other words, after an introduction pertaining to gene mutations, this chapter describes the sequential phases of wound healing, while each phase is dealt with in more detail in the subsections that follow.

Applied mathematicians can take advantage of this chapter as it offers a painless introduction to a field conceptually very far from mathematics, while biologists, may be disappointed for a not sufficient description, may look at the descriptions presented in the following simply focusing on biological phenomena that may possibly be constrained into mathematical equations. Indeed, the aim of this chapter consists in offering a brief description of phenomena that will be modeled in the third part of this monograph.

Finally, the last part of the chapter proposes a critical analysis concerning the crucial problem of reducing complexity by means of the design of a new systems biology approach, this is the main conceptual output of the first part of this monograph, which will be dealt with in the next chapter.

4.2 Genes and Mutations

DNA (Deoxyribose-Nucleic-Acid) consists of a chain made from four types of nucleotide (*bases*) subunits: *adenine, cytosine, guanine, and thymine*. Each nucleotide subunit consists of three components: a phosphate group, a deoxyribose sugar ring, and a nucleus-base. The most common form of DNA in a cell is in a double helix structure, in which two individual DNA strands twist around each other in a right-handed spiral. In this structure, the base pairing rules specify that guanine pairs with cytosine and adenine pairs with thymine. The base pairing between guanine and cytosine forms three hydrogen bonds, whereas the base pairing between adenine and thymine forms two hydrogen bonds. The two strands in a double helix must therefore be complementary, that is, their bases must align such that the adenine of one strand are paired with the thymine of the other strand, and so on.

The expression of genes encoded in DNA begins by transcription of the gene into RNA (Ribose-Nucleic-Acid), which is a second type of nucleic acid that is very similar to DNA, but whose monomers contain the sugar ribose rather than deoxyribose. RNA also contains the base *uracil* in place of thymine. RNA molecules are less stable than DNA and are typically single-stranded.

A *gene* can be defined as a region of DNA that controls a hereditary characteristic. It usually corresponds to a sequence used in the production of a specific protein or RNA (see Figure 4.1). Genes that encode proteins are composed of a series of three-nucleotide sequences called *codons*, which serve as the words in the genetic language. The genetic code specifies the correspondence during protein translation between codons and amino acids. The genetic code is nearly the same for all known organisms.

Accordingly with the recent development, the process of producing a biologically functional molecule of either RNA or protein is called *gene expression*, and the resulting molecule is known as *gene product*, and is responsible for the development and functioning of all living things. Thus a gene is a portion of DNA that contains both "coding" sequences that determine what the gene does, and "non-coding" sequences that determine when the gene is active (expressed). When a gene is active, the coding and non-coding sequences are copied in a process called *transcription*, producing an RNA copy of the gene's information. This piece of RNA can subsequently direct the synthesis of proteins via the genetic code. In other cases, the RNA is used directly, for example as part of the ribosome.

The total complement of genes in an organism or cell is known as its *genome*, which may be stored on one or more *chromosomes* (a single, very long DNA helix on which thousands of genes are encoded). The region of the chromosome at which a particular gene is located is called its *locus*.

Bearing all the above in mind, a concise definition of a *gene*, based on complex patterns of regulation and transcription, genic conservation and non-coding RNA genes, has been proposed in [Gerstein *et al.* (2007)]:

The physical observable and behavioral characteristics of an organism, called **phenotype***, can be thought as the output of a product of genes interacting with each other and with the environment. Variations in phenotype are due to variations in genotype, or the organism's particular set of genes, each of which specifies a particular trait. All organisms have many genes corresponding to many different biological traits, some of which are immediately visible, such as eye color or number of limbs, and some of which are not, such as blood type or increased risk for specific diseases, or the thousands of basic biochemical processes that comprise life.*

Different forms of a gene, which may give rise to different phenotypes, are known as *alleles*. Organisms such as the pea plants Gregor Mendel (1822-1884) worked on, along with many plants and animals, have two alleles for each trait, one inherited from each parent. Alleles may be dominant or recessive; dominant alleles give rise to their corresponding phenotypes when paired with any other allele for the same trait, whereas recessive alleles give rise to their corresponding phenotype only when paired with another copy of the same allele. For example, if the allele specifying tall stems in pea plants is dominant over the allele specifying short stems, then pea plants that inherit one tall allele from one parent and one short allele from the other parent will also have tall stems. Mendel's work found that alleles assort independently in the production of gametes, or germ cells, ensuring variation in the next generation.

The estimate of the number of gene in humans has decreased as our knowledge has increased. As of 2001, humans are thought to have between 30,000 and 40,000 genes.

A gene carries biological information in a form that must be copied and transmitted from each cell to all its progeny. This includes the entire functional unit: coding DNA sequences, non-coding regulatory DNA sequences, and introns.

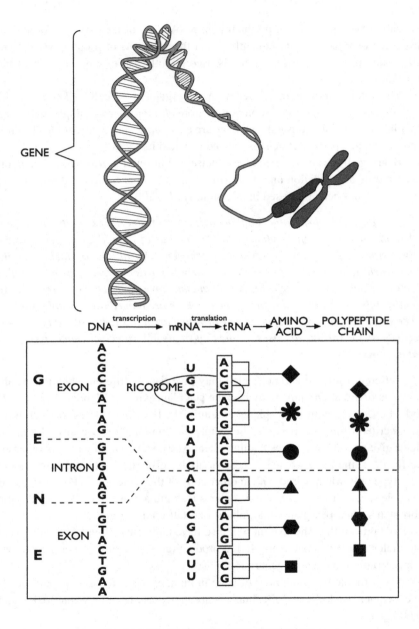

Fig. 4.1   The structure of a gene and its products.

DNA replication is for the most part extremely accurate, with an error rate per site of around 10^{-6} to 10^{-10} in eukaryotes [Watson *et al.* (2004)].

Rare, spontaneous alterations in the base sequence of a particular gene arise from a number of sources, such as errors in DNA replication and the aftermath of DNA damage. These errors are called *mutations*. The cell contains many DNA repair mechanisms for preventing mutations and maintaining the integrity of the genome. However, in some cases such as breaks in both DNA strands of a chromosome, repairing the physical damage to the molecule is a higher priority than producing an exact copy. Due to the degeneracy of the genetic code, some mutations in protein-coding genes are silent, or produce no change in the amino acid sequence of the protein for which they code. Mutations that do have phenotypic effects are most often neutral or deleterious to the organism, but sometimes they confer benefits to the organism's fitness. Mutations range in size from a single DNA building block (DNA base) to a large segment of a chromosome and propagate to the next generation lead to variations within a species' population. Differences in alleles may give rise to differences in traits. Although it is rare for the variants in a single gene to have clearly distinguishable phenotypic effects, certain well-defined traits are in fact controlled by single genetic loci. A gene's most common allele is called the wild type allele, and rare alleles are called mutants. However, this does not imply that the wild-type allele is the ancestor from which the mutants are descended.

Gene mutations occur in two ways: they can be inherited from a parent or acquired during a person's lifetime. Mutations that are passed from parent to child are called *hereditary mutations* or *germline mutations* (because they are present in the egg and sperm cells, which are also called germ cells). This type of mutation is present throughout a person's life in virtually every cell in the body.

Acquired (or *somatic*) mutations occur in the DNA of individual cells at some time during a person's life and can be caused by exposure to mutagens (ultraviolet or ionizing radiation, mutagenic chemicals, or viruses).

The corruption of a genetic locus (a specific site along the length of a specific chromosome) may occur because of physical or chemical exogenous agents, or, more usually, because of some mistake during DNA replication, in the process of the mitosis of the cell. Remembering that approximately 10^{16} mitoses occur during a normal human life span, it seems likely that every day everyone has to suffer some kind of genome alteration, despite the protection of stem cells and the efficient processes of "check and repair" of the duplicated DNA, which are typical of the mammalian organisms. Acquired mutations in somatic cells (cells other than sperm and egg cells) cannot be passed to the next generation.

Luckily, not every genetic corruption leads to the arising of a tumor: many of them are indifferent to the normal cell life, many others make the genome of the mutated cell so unstable that the cell dies after a short time. It is supposed that more than four successive genetic mistakes are necessary for a cell to become malignant. Moreover, the vast majority of mutated cells remain in a dormant, pre-malignant state for an entire lifetime.

It is worth stressing that the number of genetic mutations which are required in order to justify tumorigenesis is very high, far beyond the number of genetic mutations which occur during normal human life. However, sometimes a random DNA mutation may provide a normal cell with a sort of *genomic instability*: even if it does not seem to provide an immediate benefit in terms of proliferation and survival, genomic instability greatly increases the speed of further genetic mutations and makes easier the acquisition of other characteristics. Even if not easily quantified, the acquired genomic instability should be considered as the first hallmark in tumorigenesis.

Therefore a cell may incur into a specific alteration in one of its genes, which gives to it a significant advantage with regard to survival and proliferation, and allows it and its descendent to quickly advance along the tumor progression. Only a small portion of hundreds of the alterations present in many neoplastic cell's genome actually play causal roles in the process of tumorigenesis: mutation of a small number of critical control genes seems to be the way to acquire neoplastic cell phenotype. Generally, a tumor develops *progressively*, demonstrating different gradations of abnormality along the way from being benign to metastatic. Besides, a tumor generally arises from the genetic mutation of a single cell (the so-called *monoclonal cancer*, but there are, at the same time, descendant cells with different types of genetic mutations, i.e., different progression; thus the tumor tissue presents *genetic heterogeneity* of cells. Between two extremes of fully normal tissues and highly malignant tissues lies a broad spectrum of intermediate appearance, which we will read as a statistical distribution heterogeneity of cells progression. The different gradations of abnormality may well reflect cell populations that are evolving progressively away from normal and toward greater degrees of aggressive and invasive behavior.

Human cancer cells share a set of common characteristics (the so-called *hallmarks of cancer*), acquired on the way to full malignant state, as documented in the seminal paper by Hanahan and Weinberg [Hanahan and Weinberg (2000)]:

(1) autocrine signalling, namely capability of providing their own growth signals;

(2) Resistance to growth-inhibitory signals;

(3) Capability to proliferate indefinitely;

(4) Capability of angiogenesis;

(5) Capability of escaping apoptosis;

(6) Metastatic capability;

(7) Capability of avoiding the immune system attack.

Mutations can also occur in a single cell within an early embryo. As all the cells divide during growth and development, the individual has some of the cells, which undergo a mutation and some cells without a genetic change. This situation is called *mosaicism*.

Some genetic changes are very rare; others are common in the population. Genetic changes that occur in more than 1 percent of the population are called *polymorphisms*. They are common enough to be considered a normal variation in the DNA. Polymorphisms are responsible for many of the normal differences between people such as eye color, hair color, and blood type. Although many polymorphisms have no negative effects on a person's health, some of these variations may influence the risk of developing certain disorders. Indeed, it is well established that almost all tumors arise because of random DNA mutations which corrupt the genes of a cell, and alter the genetic regulatory circuits which control the cell functions. Therefore, normal cells evolve progressively through increasingly abnormal (dysplastic) states until the neoplasia. The formation of a tumor and its development, the so-called **tumor progression**, is quite a complex process that normally spreads over many years (ten or more).

4.3 The Phases of Wound Healing

The human body, during life, continually undergoes processes of injury, reaction, and recovery. A wound can be defined as a rupture of the natural cohesion of tissues, with the tendency to heal. *Wound healing*, or *wound repair*, is an intricate process by which the skin, or some other organ, repairs itself after injury. In normal skin, the *epidermis* (*outermost layer*) and *dermis* (*inner or deeper layer*) exist in a steady-stated equilibrium, forming a protective barrier against the external environment. Once the protective barrier is broken, the normal (physiologic) process of wound healing is immediately put in motion. The tendency to heal belongs to the wound. The healthy body reacts to harmful influences with a healing process through which a new integrity is established and the damage is repaired (*salutogenesis*).

The approach of systems biology leads to examine physiological and pathological processes in terms of the processes themselves. Collected data from the field of natural science can, with the aid of phenomenological observations, be classified and interpreted. Focusing on wound healing, the process results characterized by four sequential, partially overlapping, phases (see Figure 4.2): *hemostasis* (that is not considered a phase by some authors), *inflammatory*, *proliferative*, and *maturation (remodeling)*.

Although these phases overlap in time and interact with the other stages, they are distinct enough in their function to be considered as separated. Every phase of the healing process consists of complex interactions between cells and mediators which tend to regulate the process. The extreme complexity of these processes demands perfect coordination. It should be clear here that their coordination must be understood from the organizational level of the organism as a whole and not just as the result of a variety of cells and mediators.

The sequence of events that occur in an organism after an injury form a closed cycle. The sequence of hemostasis, inflammation, proliferation, and maturation is an essential condition for the successful completion of the healing process. The cycle of healing and repair is self-regulating. However, the process is susceptible to interruption or failure leading to the formation of chronic non-healing wounds. Factors that may contribute to such a process include diabetes, venous or arterial disease, old age, and infection.

In general, wound healing is comparable for the various types of tissues. The wound healing process of the skin can be observed directly and is immediately accessible to study. It has, therefore, been considered as a model for the illustration of the general principles of wound healing in every tissue.

An insight into the processes in an organ of repair can be provided in the following manner:

Each phase lasts increasingly longer than the previous one;

Hemostasis takes anywhere from a couple of minutes to an hour;

Inflammation takes days, proliferation takes months, and maturation takes up to approximately a year after the injury;

The phases also flow into each other and exist, in part, simultaneously.

The various phases are represented in Figure 4.2, while their overlap is visualized in Figure 4.3.

4.3.1 *Hemostasis Phase*

The appearance of blood is the immediately first sign of a wound. If there is no arterial hemorrhaging involved, the bleeding will stop spontaneously after a fairly

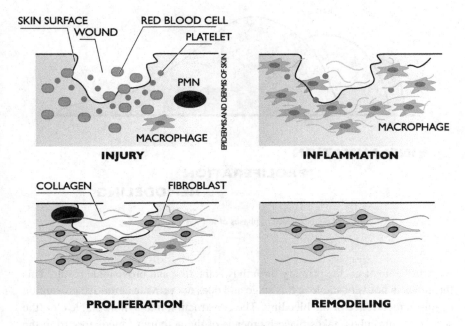

Fig. 4.2   The phases of cutaneous wound healing.

short time. The primary reaction to the injury is to control the blood loss. Stopping of the bleeding is called *hemostasis*. During the hemostasis we can distinguish between three different types of reactions:

• *Vascular reaction*. The smaller damaged blood vessels (arterioles) contract reflexively in two directions: by drawing back into the tissue (retraction) and by contraction of the blood vessel itself which closes the vessel (reflexive vasoconstriction);

• *Cellular reaction*. When the inner lining of the blood vessel (endothelium) becomes damaged, that results into the exposure of the connective tissue (collagen) around it to the circulating blood (collagen exposure). Part of the collagen, the extra-cellular matrix, activates the thrombocytes in the blood and is a strong stimulator of coagulation (thrombogenic). Subsequently, the coagulation cascade is triggered in such a way that a whole series of mediators are activated in a specific order. Together with the activated thrombocytes, this ultimately leads to the forming of a blood clot (thrombus) in the severed end of the blood vessel. At this

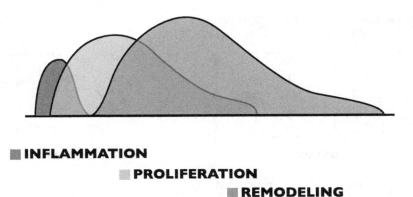

■ INFLAMMATION

        ■ PROLIFERATION

                ■ REMODELING

Fig. 4.3   Sequence of phases of normal wound healing.

stage of hemostasis, the primary thrombus is the first and provisional result. This thrombus in not very stable at this stage and does not provide definitive prevention against a recurrence of the bleeding. The construction into a more stable clot (the secondary thrombus) takes place by means of the activity of mediators from the blood serum.

• *Humoral reaction and production of mediators.* The activation of thrombocytes results in the release of more mediators from their granulae which further support the vasoconstriction. Alongside of that, an immediate production of thromboxane occurs that is secreted by the thrombocyte with the aid of the Cox-1 enzyme. The outcome is a further, humorally mediated, vasoconstriction. Further reactions of the thrombocytes stimulate the release of mediators of the co-agulation cascade. The activated coagulation cascade in the blood serum ensures stabilization of the primary thrombus. Because of that, a secondary, fixed clots develops with the aid of thrombocyte contraction and fibrin forming.

    The secondary thrombus forms a primary substance for the matrix, the basic structure within which the further processes of wound healing occur. Hemostasis can be compared with the acute phase reaction of the innate immune system during infection. The acute phase reaction also results in the provisional recovery of integrity. In the case of the bleeding wound and for the immune reaction a whole range of physiological reactions is to follow, that ultimately leads to definitive recovery of the integrity. It is worth stressing that the hemostasis develops locally.

4.3.2   Inflammation Phase

The inflammatory phase begins from an hour to a day after the injury and runs until the commencement of the proliferation phase.

The central task is the phagocytosis of undesirable material, such as cell debris and microorganisms. The removal of tissue is characteristic for the beginning of the process. The organism literally clears the path for the growth of new tissue.

This phase is actually the first active reaction of the organism as a whole to the tissue damage that has occurred. The inflammation phase of the wound healing process will actually appear to be characteristic of a complete activation of the organism by itself.

Just as in hemostasis, the inflammation is organized with the aid of a vascular, a cellular, and a humoral component.

• *Vascular reaction.* With the aid of enzymes in the endothelium, the secretion of various mediators occurs, which plays an active role in the blood vessel dilatation (vasodilation) and induce an increased permeability of the vascular wall. This is the cause of both the redness (rubor) and the warmth (calor) that we can observe macroscopically. Moreover, the vasodilation also results in a slowing of the blood flow, which ensures that new cells and mediators can reach the entire wound region.

• *Cellular reaction.* Leukocytes (in particular neutrophiles) and macrophages find their way locally through chemical attraction (chemotaxis). They phagocytose micro-organisms (bacteria, viruses, fungi, etc.), and cell remnants of the tissue. The macrophage phagocytoses all of the cells that have become apoptotic (lymphocytes as well as tissue cells) as well as other infectious cells that must be removed.

• *Humoral reaction.* Toxins and mediators are spread throughout the entire body. The person who is wounded can have fever, feel sick, have a loss of appetite.

It is worth stressing that when micro organisms that can multiply are not present, then the wound becomes infected. This is an important cause of additional tissue damage. When the wound is infected, there is an enhanced inflammation phase. The activated inflammatory process is damaging and can divert into tissue destruction or allergic reactions. The transition from acute to chronic inflammation occurs when the acute inflammatory reaction cannot be successfully completed. This may be the result of a persistent stimulus or through a disturbance of the normal healing process. Examples of illnesses in which chronic inflammation play a role are: rheumatoid arthritis, arteriosclerosis, tuberculosis, and chronic lung diseases. The characteristic of these diseases is that there is, at the same time, active inflammation and tissue destruction, alongside healing attempts in the form of fibrosis.

4.3.3   *Proliferation Phase*

After removal of the damaged tissue and elimination of a possibly present infectious pathogen, the genuine wound healing process can begin by means of proliferation of new tissue. In the proliferation phase, we can also distinguish among vascular, cellular, and humoral components.

• *Vascular reaction.* The existing blood vessels ensure the formation of new offshoots, a process that is called neo-vascularization (angiogenesis). The development of new blood vessels progresses via a number of steps: migration of the inner lining of the blood vessel (endothelial cells) in the direction of the stimulus that generates angiogenesis; increase of endothelial cells, just behind the front of the migrating cells; maturation and remodeling of endothelial cells; recruiting of peri-endothelial cells: cells that form the tissue surrounding the endothelium.

• *Cellular reaction.* During the proliferation phase, cell replication takes place such that new tissue develops, the so-called wound-healing matrix. Characteristic for this phase is the formation of new tissue. The fibroblast, a specific connective tissue cell, plays a role in the formation of connective tissue (fibroplasia), the development of granulation tissue, and matrix formation; the epithelial cell (keratinocyte) of the epidermis plays a role in covering the wound (epithelization). The macrophage continues to play an important role in regulating the process and in taking on the functions that the leukocyte fulfilled during the inflammation phase.

• *Humoral reaction.* This reaction progresses with the aid of several humoral factors for processes such as further matrix formation out of the secondary thrombus, and epithelization. The macrophage has, once again, an initiating role through the secretion of various signaling substances and growth factors.

It is worth mentioning that the tissue recovery consists of the combination of two processes depending upon the degree of the damage:

a) *Regeneration* (normal tissue): Tissue recovery dominates in regeneration through proliferation of the parenchyma cells that are already present. The precondition for regeneration is the presence of an intact basal membrane, which functions as an underlying and supportive framework. Regeneration generally leaves no macroscopic sign of injury.

b) *Fibrosis* (scar tissue): The breakdown of normal tissue dominates in fibrosis through the collagen secretion of fibroblasts. Destruction of the basal membrane on top of the destruction of parenchyma cells leads to fibrosis. This results into a permanently visible scar.

Both processes make use of comparable biological mechanisms, which include cell proliferation, cell differentiation, and interactions between cell and matrix. The proliferation phase is, just as hemostasis and inflammation, a process in equilibrium. The disturbed balance of proliferation can diverge to the not-actively healing wound in the form of the ulcus, or to excessive production of wound tissue, the granuloma.

4.3.4 *Maturation or Remodeling Phase*

Improvement of the collagen in the scar to the greatest possible perfection is characteristic for the remodeling. This development determines the transition of a temporary matrix to a definitive matrix. In the collagen, a remodeling takes place with a continuous resorption and re-deposition of collagen, changing the structure and the quality of the matrix. The process moves forward from a temporary structure to a definitive structure, from temporary to definitive tissue, and from an orientation of the collagen fibers parallel to the wound edges to an orientation along lines along which mechanical load occurs, the so-called stress lines. The mechanical forces working on the tissue determine the ultimate form it will take. In this phase, we can differentiate among vascular, cellular, and mechanical reactions.

• *Vascular reaction.* The blood vessels from the granulation tissue regress further. For the duration of the transformation of scar tissue, there is, therefore, a generalized reduction of blood vessels, so that, ultimately, there is a white scar in which there are no more blood vessels present.

• *Cellular reaction.* The proliferation supported by collagen synthesis has stopped and the transformation of the scar begins under the influence of physical-mechanical stretch and strain. Fibroblast apoptosis occurs resulting in a normal looking scar.

• *Mechanical reaction.* In this phase, humoral factors play a subordinate role. Their function is assumed by the mechanical stretch and strain that is exerted on the tissue in daily life. These now provide the impulse for the final transition.

During tissue remodeling, the equilibrium now shifts in favor of tissue breakdown and apoptosis. The modeling of tissue in this phase is primarily determined through mechanical forces and no longer through growth.

It is worth stressing that, when tissue production remains dominant over apoptosis, an increased collagen synthesis will lead to a hypertrophic scar or *keloid*. A hard, thick lump of overgrown tissue develops as the result of insufficient resorption and apoptosis. Here, the original form is not optimally restored as well. The abnormal form of the scar remains due to a hardening tendency.

4.4   The Fibrosis Disease

This section is focused on the pathological aspects of the fibrosis phenomenon, that is on the effects due to the so-called fibro-proliferative diseases. From the mathematical point of view this subject has been widely studied at macroscopic level, but insight at cellular and sub-cellular level scale is still lacking [Fusi (2009)]. Therefore, there is a remarkable need of developing the necessary mathematical and computational approach that can handle the full range of time and length scales required to model complex biological systems, with emphasis on spatially distributed systems, and apply this framework to the fibrosis process.

As expressed in the previous section, when the normal wound healing response fails, the formation of scar occurs very fast and production and deposition of collagen results in a pathological scarring, called it fibrosis, which is implicated in the onset of a number of cancers.

Organs, in the fibrotic process, become stiff and cannot perform functions essential to life and health, leading to organ failure and death. Fibrosis can be triggered by a variety of events including trauma, surgery, infection, environmental pollutants, alcohol, and other types of toxins and it can potentially involve most tissues and organs of the human body. The fibrotic progression is characterized by the termination of the regular organ repair functionality and the development of fibro-proliferative wound healing. This type of abnormal healing can be regarded as pathologically excessive responses to wounding in terms of cells profiles and their inflammatory growth factor mediators.

Phenomenologically, fibro-proliferative diseases commonly exhibit increased healing responses and this abnormality results in a combination of chronic inflammation, fibro-proliferative and non-regenerative repair. The fibro-proliferative process leads to different pathologies, involving every organ of the human body and typically found in elder people (more than 50 years old).

Fibrosis represents the repair of a damaged tissue where parenchymal cells are replaced by connective tissue. When this occurs, the healing process cannot evolve through the regeneration of parenchymal cells and deposition of connective tissue leads to the formation of scars. Detailed biological insight into the mechanisms of pathogenesis, progression, stabilization and regression of fibrosis is lacking, and the existing clinical corrective methods are typically long-term, unpredictable, traumatic for the patient and prone to failure or recurrence. Unlike inflammation, for which anti-inflammatory therapies abound, there are no approved treatments that directly target the process of fibrosis despite its preponderance and potentially fatal consequences in so many diseases. Current therapies for fibro-proliferative disorders usually include anti-inflammatory drugs, which are palliative at best and fail to address the fibrotic process that causes disease progression. There is a large unmet need for a safe and effective anti-fibrotic therapy that delays diseases progression and reduces mortality.

Below we have outlined some types of fibro-proliferative diseases, briefly describing the evolution of the pathology. Of course it is not an exhaustive collection, but it gives the reader an idea of the main features that should be taken into account when developing mathematical models for fibro-proliferative pathologies.

• **_Liver Fibrosis._** Liver fibrosis is a fibro-proliferative disease where the pathological wound healing response results in hepatocytes necrosis and apoptosis. In liver fibrosis wound healing is mediated by some cells called myofibroblasts (see [Iredale (1998)]). After injury, some resident hepatic stellate cells (HSC) undergo myfibroblasts-like phenotype conversion, which may mark the onset of fibrosis. This transformation is characterized by expressions of profibrotic genes like collagen 1, MMPs (matrix metalloproteinases) and TIMPs (tissue inhibitor metalloproteinases). In the normal healing process, after phase (iii), in which extracellular matrix is produced to avoid the mechanical collapse of the tissue, the collagen, which is the main component of the extracellular matrix, goes to progressive degradation to restore the pre-wounded state. In liver fibrosis MMPs, which are responsible of extracellular matrix degradation, are scarcely expressed and the collagen production can be up to five times larger than in normal healing response. In this pathological condition the production of TIMPs (inhibitors of MMPs) grows exponentially, as HSC differentiate into myofibroblasts, creating a "chronic" condition in which the extracellular matrix is continuously produced. This suggests that the balance between MMPs and TIMPs, which regulates the production/degradation of the extracellular matrix, is the key parameter of this process and indicates that liver fibrosis is a potentially reversible process (see [Pinzani (2000)]).

• **_Dermal wound healing._** This is probably the largest class of fibroproliferative disorders that has been studied from the mathematical point of view. The healing of a skin wound involves a series of repair process which can depend upon species, age, wound dimension and other factors such as infections and anatomical location, see [Clark (1989)]. Moreover foetal and adult wound healing is substantially different (see [McCallion and Ferguson (1996)]). Adult mammalian skin is composed of two layers called the *epidermis* and the *dermis*, the first being the outermost. The epidermis is composed of keratinocyte cells which proliferate and differentiate (see [Dover and Wright (1991)]). The dermis is composed by fibroblast cells, extracellular matrix (ECM), blood capillaries and other structural components (see [Odland (1991)]). The extracellular matrix is made of diverse functional macromolecules (proteins and collagens) and constitutes an essential regulator of cell physiology. It is implied in morphogenesis, cell survival, cell cycle, cell migration and tumorigenesis (see [Basbaum and Werb (1996)]). Beneath the dermis, substrata of fatty and fibrous tissue, together with blood vessels are found (see [Mast (1993)]). The epidermal response to wounds has been extensively studied and various mathematical models are found in the literature (see e.g. [Sherratt, Martin, and Lewis (1992); Sherratt and Murray (1990, 1992)]).

In dermal wound healing, several overlapping phases are initiated in order to complete the repair process: *inflammation, proliferation* (consisting in tissue formation, angiogenesis and tissue contraction) and *remodeling* (see [Clark (1989)], [Kirsner and Eaglstein (1993)]). During the inflammatory phase platelets start to migrate from damaged blood vessels forming a blood clot. They seal the wound, segregate a series of biochemical substances (such as growth factors) and trigger some enzymatic processes that initiate the healing phase of the wound (see [Martin and Lewis (1992)]).

The proliferative phase usually starts some days post-wound. In this phase angiogenesis takes place and blood supply together with metabolic activity is established in the wound site. Moreover, some growth factors chemotactically recruit fibroblasts into the wound, where these cells proliferate to produce growth factors and ECM. This set of processes usually go under the name of *fibroplasia*. Some of the fibroblasts are converted into myofibroblasts which possesses molecular and structural components typical of contractive cells. Myofibroblasts are responsible for wound contraction. The movement of the fibroblast into the wound site produces traction forces on collagen fibres. The tension exerted is transmitted through the wound by myofibroblasts (see [Singer *et al.* (1984)]) and this culminates in wound contraction.

After many months fibroplasia ends, the ECM starts to remodel. Inflammatory cells and myofibroblasts are not present in the wound site any longer, but fibroblasts continue to remodel the ECM in order to restore the mechanical and biological properties of the original tissue. This is known as the *remodeling* phase.

• *Corneal wound healing.* Corneal wound healing is quite different from all other mechanisms encountered in other organs (see [Forrester (1996)]). For instance, in non-pathological conditions, it does not entail angiogenesis, which is a peculiarity of the inflammatory phase of the vast majority of tissues. The structure of the cornea is made of a transparent epithelial layer covered on top by a liquid tear film. The epithelial layer contains some growth factors among which we can find the epidermal growth factor (EGF), that seems to be the main chemical which regulates the balance between cell loss and proliferation (see [Gipson (1992); Gipson and Inatomi (1995)]).

The corneal wound healing is a multi-phase process. After the cornea has been injured, the exposure of the epithelial layer to the tear fluid results in the expression of fibronectin (a glycoprotein binding extracellular matrix components such as collagen and fibrin) in the wound site. After this, a monolayer of epithelial cells starts to spread in the wound bed, forming and breaking focal contacts (adhesions by which cell attach to the underlying substrate) with the expressed fibronectin. This process is followed by the migration of a multilayer of undifferentiated epithelial cells to fill the epithelial defect. In the final phase the epithelial cells differentiate to regain the pre-wounded original structure. Epithelial corneal fibrosis or haze, is the pathological counterpart of the corneal wound healing. On a molecular level, there is a production of disorganized collagen as well as an increase in cellular density that is responsible for haze formation.

• *Bone wound healing.* Bone, like other tissues, has the capability of repairing itself. The fracture healing process can be essentially divided into three parts.

i) Reactive phase: after fracture, blood cells start to form clot in the wound site and blood vessels constrict to prevent bleeding. In the region adjacent to the wound site fibroblasts form aggregates of cells known as granulation tissue (see [Ham and Harris (1972)]).

ii) Reparative phase: after some days the cells of the periosteum (exterior part of the bone) near the bone gap develop in chondroblasts and form the so-called *hyaline cartilage*. Other periosteal cells distal to the fracture develop into osteoblasts and form *woven bone* (weak structure with high number of osteocytes). Also the fibroblasts in the granulation tissue help forming the hyaline cartilage. The hyaline cartilage and the woven bone grow in size until they unite to form the fracture callus ([Brighton and Hunt (1986)]). The subsequent step is the so-called endochondral ossification, where hyaline cartilage and the woven bone are substituted by lamellar bone. This process occurs after the collagen matrix of either tissue becomes mineralized. New lamellar bone, called trabecular bone, is then formed upon the recently exposed surface of the mineralized matrix. Eventually, all of the woven bone and cartilage of the original fracture callus is replaced by trabecular bone, restoring the bone's original strength.

iii) Remodeling phase: this is the phase when trabecular bone is substituted with compact bone. The trabecular bone is first resorbed by osteoclasts, which subsequently deposit compact bone. Eventually, the fracture callus is remodeled into a new shape which closely duplicates the bone's original shape and strength.

In the healing process a key factor is the growth factor secreted by the osteoblasts. Among these the one that affects bone growth is the bone morphogenetic protein (BMP). Some models take into account the so-called critical size defect (CSD), which is defined as the diameter of a bone wound beyond which complete calcification of the wound will not occur during lifetime.

4.5   Critical Analysis

The phenomenological description proposed in this chapter has been focused on a variety of biological processes. The modeling approach should be demanded to describe them by equations that involve the interactions with the outer environment as well as the application of therapeutical actions.

It is not an easy task due to the great complexity of the biological phenomena we deal with. An important source of complexity is the multiscale characteristic of all phenomena from the scale of genes to that of tissue. Namely, the dynamics of cells is ruled at the low molecular scale that determines their phenotypes, while aggregation of cells generate the dynamics of tissues. The physical behavior of tissues evolves on time due to mutations of genes' expression.

Different classes of mathematical equations are needed to model the system at different scales. Moreover, the mathematical approach is required to treat the passage from one scale to the other. These conceptual and technical difficulties suggest to reduce complexity by a new system biology vision. This need has already been anticipated in this chapter and will be specifically developed in Chapter 5.

Chapter 5

From Levels of Biological Organization to System Biology

5.1 Introduction

The onset and evolution of complex biological phenomena, such as healing processes, need a deep analysis of the different functions expressed by the interacting biological entities that compose the system, at each level of their activity or scale: genetic, cellular and tissues. Generally, all these scales are needed to represent a real biological system through mathematical equations. Therefore, the idea straightforwardly transferring a whole variety of phenomena into models is surely naive in the absence of a constructive effort to reduce complexity.

The modeling approach needs to reduce system biology approach to reduce the complexity of large systems by decomposing them into several interacting subsystems, where each subsystem is characterized by the biological function it expresses. The theory of modules proposed by Hartwell [Hartwell *et al.* (1999)] can be used to deal with this problem. Focusing on the cellular scale, a constructive interpretation of the modules theory [Bellomo and Forni (2008)] suggests considering the collective behavior of systems of cells which have the ability to express certain biological functions.

The natural development of the theory of modules consists in identifying each module as a functional subsystem where different entities and components have the ability to express one specific biological function, which is only identified by a scalar variable. The interactions among various subsystems play a relevant role on the overall dynamics. Hence, this system is a network (system of systems) of interacting sub-systems each of which has been defined at different scales.

It is worth stressing that a subsystem is an entity which has to be defined with reference to the specific analysis under consideration. In other words, the same system can be decomposed in different ways according to the modeling purpose. Therefore, the first step of the modeling approach consists in understanding how the overall system can be decomposed into functional subsystems on the basis of the specific biological investigation under consideration. This concept, which has already been anticipated in the previous chapters, can be made less painful to follow if it is properly focused, as we shall see, on specific examples.

This chapter deals with a personal quest into the methodological approach of systems biology which involves the concept of functional subsystems and scaling. Subsequently, the modeling of the components of the overall system, namely of the functional subsystems, is developed at each specific scale are related to the appropriate class of equations. Finally some preliminary reasonings on the links between scales complete this chapter. The contents aim at offering the conceptual background to the modeling methods proposed in the next chapters.

5.2 From Scaling to Mathematical Structures

Let us consider the assessment of the *scaling problems* referred to the concept of functional subsystems. Moreover, let us preliminary define the concept of *multi-scale* in the context of modeling biological systems, namely by clarifying the meaning of scale in a biological or physiological sense. A natural way to achieve this target in space consists in splitting the processes according to their position within the biological hierarchy, i.e., whether they represent interactions between organs, within a tissue, between cells, and so on. We can also refer to these hierarchical positions in terms of levels of biological organization.

Bearing all above in mind, let us describe the relevant biological phenomena at each scale that need to be considered in the modeling approach.

The *molecular scale* (or *genetic scale*) refers to the events which regulate the evolution of a cell. Genes, contained in the nucleus of the cell, are responsible of the activities expressed by cells. Receptors on the cell surface can receive signals that are then transmitted to the cell nucleus, where the aforementioned genes can be activated or suppressed. In extreme situations, particular signals can induce uncontrolled cell proliferation, or cell death- so-called apoptosis or programmed cell death. For instance, unregulated proliferation may activate interactions between tumor cells and host cells, which occur at the cellular level, but are mediated by subcellular processes, such as through signal cascades and receptor expression.

These interactions can result in temporary, or even permanent, alterations in gene expression, which in turn can affect a cell's state, such as activation or inactivation of immune cells. Additional concepts on genes and genomic sciences is given in the next section. The structure of the cell and the dynamics of gene expression are represented in Figure 5.1.

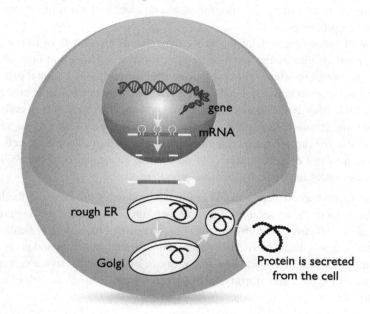

Fig. 5.1   Dynamics of gene expression inside a cell.

The important phenomena related to the *cellular scale* correspond to the cell-cell interactions, which are key elements at all stages of biological events (healing process, fibrosis, tumor formation). The typical size of a cell (about 10^{-2} millimeters, despite a great deal of variety between different types of cell) also puts them at the interface between the mostly biochemical processes at the genetic level and the resulting macroscopic physiological properties.

Cells are also self-contained and isolated from their surroundings by a selectively permeable membrane (and in the case of prokaryotic cells also by a rigid cell wall). Therefore, many of the interactions of a cell with other cells and the extracellular space can be classified by recording the flux of various substances (ions, nutrients) across the cell membrane.

For instance, cell-cell interactions have a key role in the early tumor development. If the immune system is active and able to recognize tumor cells, it may be able to develop a destruction mechanism and induce cancer cell death; otherwise, the tumor may evade apoptosis or co-opt the host cells, allowing progressive growth. During invasion and metastasis, alterations in cell-cell adhesion between individual tumor cells are key to driving the process. These and other cellular interactions are regulated by signals emitted and received by cells through complex transduction processes.

A well-known example of the above dynamics is the growth of tumor cells, which, if not cleared by the immune system, forms a mass so that macroscopic features become important. However, even after formation of a tumor structure, interactions between individual cells are crucial to driving macroscopic processes.

An analogous dynamics occurs in wound healing and repair, where cells proliferate to reach to a pathological state, subsequently the system may regress to a normal state or further degenerate into proliferation of pre-neoplastic cells. Interactions may not be proliferative, but may cause modification of the ability to express biological functions by learning processes.

The *tissue scale* represents the ultimate product of the events at the lower scale, for instance the tumor mass or cancer metastasis. In practice, cells aggregate into condensed forms. In addition to the cells, a region of tissue will typically also contain an extra-cellular matrix (ECM) that holds the cells together and gives them some structural stability. Wound healing processes can generate tissue corruption and repair. In some cases, formation of special structures, such as keloid tissue, are generated.

These reasonings on the scaling problem do not claim to be exhaustive, but simply to provide a natural range of length scales in many applications. For instance, in tumorigenesis (the formation of a cancer, thought to be due to a genetic mutation) an appropriate initial choice of levels might be: genetic (gene expression), cellular, tissue.

In some applications it may be more natural to merge two of the levels discussed here or to split one or more of them into distinct sub-levels. However, to study healing processes and in general repair processes can be opportune consider an additional scale, namely the *organ* scale.

• An *organ* is a discrete unit performing a function or group of functions (e.g. the heart, brain, lungs). Usually the organ consists of a single main tissue type (e.g. myocardium in the heart) and several sporadic types (nervous, connective, blood). At the organ scale, modeling usually consists of integrating tissue level models with one another and with a representation of the geometry of the organ that is being modeled. Various organ level heart models have been defined [Kerckhoffs *et al.* (2006)].

It is worth stressing that the way by which the model is derived should take into account the essence of the system that is modeled rather than simply follow a rigid framework set out in advance. Accordingly, the decomposition into functional subsystems can be organized following the guidelines:

- The decomposition at the molecular scale consists in the selection of groups of genes whose over and lower expression can cause undesired phenotype expressions at the cellular scale. Among others, loss of programmed death ability.
- The decomposition at the cellular scale corresponds to different cell populations that collectively have the ability to express a certain function. For instance, proliferation tumor cells corresponding to different levels of mutations, or immune cells that express the ability of targeting tumor cells and develop a destructive activity.
- The decomposition at the tissue scale consists in the identification of different tissues corresponding to different types of mutations, and more in general to the underlying dynamics at the lower scales.

The modeling approach needs the identification of the representation scale for each functional sub-system and of the interactions among functional subsystems including interactions at different scale. Such a wide range of scales requires a coupled hierarchy of models, each describing behavior at a different scale, to encapsulate the complete system.

In this context the biological system is called *multiscale*. It is worth stressing that all events at the cellular and tissue scales are governed by the dynamics at the genetic scale. Therefore, the modeling at the higher scales needs to be constantly referred to the lower scale.

Further, let us mention that different functional subsystems can be located in the same space domain where they interact. For instance, this is the case of different cell populations which interact, evolve, and compete. Alternatively, the localization, may refer to a network.

Different mathematical classes of equations correspond to each scale with the additional technical difficulty that each class should be consistent with those related to the same biological system at the lower and higher scale. We distinguish the following:

i) **Models at the molecular scale** are generally developed by *stochastic game theory* used to model gene expression and related biological functions which characterize models at the cellular scale;

ii) **Models at the scale of a cell** are generally developed by means of *ordinary differential equations* or *Boolean networks*;

iii) **Large multicellular systems** are modeled by nonlinear *integro-differential equations* similar to those of the nonlinear kinetic theory (the Boltzmann equation), or by *individual based-models*, which give rise to a large set of discrete equations, or by partial differential equations for systems with internal structures.

iv) **Macroscopic models** are formulated in terms of *systems of partial differential equations*, generally nonlinear, obtained by asymptotic limits of kinetic models for active particles, letting the intercellular distance tend to zero. Classically, the representation is delivered by averaged quantities, typically mass density and mean velocity.

It is very important that the modeling approach takes into account interactions involving all scales and the different mathematical representation at each scale as shown in Figure 5.2, which visualize the three scales interacting in a cancer tissue chosen as a paradigm supporting the above reasonings.

In some cases, for instance in the formation of capillary sprouts related to angiogenesis phenomena, the classical assumption of mathematics to derive specific models, for instance continuity of matter, are not valid any more.

Moreover, stochastic behaviors are an essential feature of the biological systems under consideration as visualized in the angiogenesis phenomena shown in the same figure.

The type of links between two scales evolves in time being related to the dynamics at the lower scale. Specifically, the dynamics at the molecular scale determines the functions expressed by cells. Consequently, their rule of interactions with the other cells, which end up with the specific structure of tissues that evolve in time. Therefore, also the decomposition into functional subsystems may have to be modified along the time evolution.

5.3   Guidelines to the Modeling Approach

The multiscale aspects of biological systems imply that the dynamics at the higher scale is influenced by the lower scales, namely the parameters of models at the high scale, cellular or tissue, have to be computed by the dynamics at the lower scale. This chapter has been mainly focused on the understanding of the dynamics at the lower scale with the aim of extracting the basic concepts of the organization of biological behaviors in view of their structuring into a new system biology approach.

Let us now consider the problem of understanding the delicate issue of the links between the various scales. Two important issues should be considered among many others. The first one consists in deriving the macroscopic tissue models from the underlying description delivered the cellular scale. It is a difficult mathematical problem that needs rather sophisticated tools that will be pre-

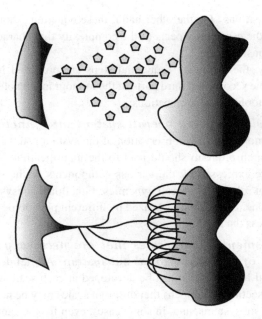

Fig. 5.2 Chemical substances, attract blood vessels, with genesis of capillary sprouts from existing vasculature to feed the tumor.

sented in the forthcoming Chapter 6 and applied to the derivation of chemotaxis macroscopic models in Chapter 9.

The second one, which has to be regarded as an open problem is the description and mathematical modeling of the interactions between functional subsystems, which need different scales. This need creates serious mathematical difficulties related to the modeling of the interactions between the scales. Suppose that a system is decomposed into a certain number of functional subsystems and one of them is, its own, a multiscale system, the question is how to describe the interactions between this functional subsystems and the others. The problem becomes extremely difficult when each of the functional subsystems has a multiscale structure.

Bearing all the above in mind, let us summarize the sequential steps to be followed in the approach to modeling complex living systems. This approach aims both at reducing complexity and at retaining, at the greater possible extent, the ten complexity features that have been reported in the first two chapters. It is a difficult compromise considering that the overall number of variables that identify biological systems are generally too large to be constrained into tractable mathematical equations. On the other hand, these equations should not lose the ability to depict the essential aspects of the complexity that characterizes the real systems under consideration.

Accordingly, the following sequential steps are proposed in view of their development in the second and third parts of the monograph devoted, respectively, to mathematical tools and to applications:

S.B.1. Subdivision of the overall system into functional subsystems: Both phenomenological observation of the system and the specific objectives of the research program should lead to the identification of the biological functions that are expressed by the various components of the system and that have an important role in the overall dynamics. Functional subsystems are an aggregate of components, even characterized by different phenotype specificity, that collectively express the same functions.

S.B.2. Assessment of the scales that are necessary to represent the dynamics of each functional subsystem: A brief description of the various biological functions that can be developed at each scale have been given in the preceding section. It happens that different scales may be necessary for different functional subsystems, and, in some cases, even for the same system. This aspect must be carefully taken into account. Moreover, the mathematical approach should consider the fact that the derivation of models of biological tissues is related to the lower cellular scale, and subsequently the organization of functional subsystem into organs needs to be treated.

S.B.3. Modeling of the dynamics of each functional subsystem: The modeling approach has to be referred to a general mathematical structure that acts as a paradigm for the derivation of specific models. Such a structure should retain, as already mentioned, the various complexity characteristics described in Chapters 1 and 2. Let us anticipate with respect to Part II, where this topic is treated, that a unified structure is proposed referring to the molecular and cellular scales, while the structure of equations of continuum mechanics is used for the larger scale of tissues. The derivation of models needs, as we shall see, the mathematical description of the interactions involving entities of the same functional subsystems, but also interactions with the contiguous scales, lower and higher.

Linear and nonlinear interactions can be considered within the framework of a complex dynamics that involves hiding and learning processes. In general, these interactions have different outputs such as modification of the biological state, proliferative and/or destructive events, transition from one functional subsystem to the other.

S.B.4. Modeling interactions among functional subsystems: Interactions between two subsystems can act in different ways, for instance as boundary conditions or as external inputs. The main difficulty is that different scales can be necessary for each subsystem. For instance the dynamics at the low scale acts as an outer input to the dynamics at the macroscopic scale. This occur also in the case of therapeutical actions that may act at different scales.

S.B.5. Validation of models: This is a very delicate issue considering that validation must be based on empirical data related to experiments. A remarkable difficulty is that experiments in vitro refer to very special situations that only partially correspond to living reality. On the other hand, living systems show a heterogeneous behavior that differentiate the response of each different individual. Despite this difficulty, a general consideration is that mathematical models need parameters, their identification should be based on empirical data. Once the characterization of the parameters of the model has been properly developed, its validation can be assessed by looking at the ability of the model to depict a variety of biological behaviors much broader than that of the experiment.

These guidelines define a very difficult research project that on one side represents a challenging objective of theoretical biology, which can be regarded as a new frontier of mathematical sciences. In fact, new mathematical tools are necessary to derive models that capture the essence of biological phenomena.

The present state-of-the-art does not yet possess appropriate methods to pursue the above objectives. On the other hand, this monograph attempts to provide the basis to pursue and apply them to well-defined case studies. Accordingly, their treatment will be given, at least partially, in the following chapters. In particular, the critical analysis proposed in the last chapter identifies the effective contribution to the development of a mathematical theory of complex systems according to the definitions given in the first chapter.

Particularly important is the validation of models, some preliminary ideas have been given in item S.B.5, the following additional considerations can be given:

i) Parameters that characterize a model should be related to well defined biological phenomena and referred to structures that are typical of the scale used for the modeling approach;

ii) The identification of parameters should be developed not simply by comparisons of prediction of the model and the output of empirical data concerning the overall behavior of the system. Experiments should specifically refer to the biological dynamics related to the parameter under consideration.

iii) Validation should be referred not only to quantitative predictions, but also to the ability of models to depict emerging behaviors that are experimentally observed. Hopefully models may drive experiments to discover emerging behaviors that are predicted by models, but not yet observed.

Finally let us anticipate, with respect to the last chapter, that the role of mathematics is not limited to reproducing experiments. An additional challenging objective consists in developing new structures, hopefully a new theory, suitable to capture the complexity of the biological phenomena. This delicate issue will be critically analyzed in the last chapter on the basis of the mathematical approach proposed in Part II and by taking advantage of the applications developed in Part III.

PART 2
Mathematical Tools

Chapter 6

Mathematical Tools and Structures

6.1 Introduction

This chapter deals with the derivation of mathematical structures that are suitable to offer a background in order to design specific models, with reference to the two low scales. Subsequently, it is shown by focusing on the phenomenological description that was illustrated in Part I of the monograph, how such a structure can be characterized at each scale. Some guidelines on modeling are proposed to model the complex dynamics of immune competition.

The mathematical approach to model living systems needs the design of appropriate mathematical structures that have the ability to capture the complexity characteristics that were pointed in Chapter 2. Subsequently, models can be elaborated to form a mathematical theory, if the phenomenological description of the living matter is substituted by a robust theory generated by deep insight into the molecular and cellular scales. Such a theory should show how the information on genotype is transferred to the cell phenotype, and how their dynamics is consequently ruled.

This chapter presents the derivation of such mathematical structures referring to the two low scales. We claim that a general common mathematical framework can be designed both at a molecular and at a cellular scale, where systems are viewed as complex systems, which interact in a nonlinear manner. The modeling, and hopefully a theory, naturally need to be specialized differently for each of them.

Therefore, this chapter proposes a constructive response to the crucial issue of selecting the appropriate mathematical structures and provides some guidelines to design specific models at each scale.

The modeling approach is also focused on immune competition with the aim of providing additional tools to treat this complex dynamics that can play a key role in the class of biological phenomena under consideration.

6.2   Mathematical Frameworks of the Kinetic Theory of Active Particles

This section deals with the assessment of the mathematical structures that can be used at the molecular and cellular scale. Specifically, we refer to the so-called *kinetic theory of active particles*, for short KTAP's theory, introduced in the book [Bellomo (2008)] and in the review paper [Bellomo, Bianca, and Delitala (2009)].

Let us first consider the mathematical representation of a large system constituted by active particles belonging to different functional subsystems. The physical microscopic state of the particles is identified by a set of variables suitable to describe their state, while the overall state of the whole system is described by a probability distribution over the microscopic state of the particles according to the following definitions:

Definition 6.2.1. The system is constituted by a large number of interacting entities whose state, called *microscopic state*, includes not only geometrical and mechanical variables, typically position and velocity, but also an additional variable called *activity*, which represents the biological functions expressed by suitable collections of active particles.

Definition 6.2.2. Active particles are subdivided into *functional subsystems* identified by the specific activity they express.

Definition 6.2.3. The description, in probability, of the overall state of the system is defined by the so-called *generalized distribution functions* f_i over position $\mathbf{x}$ and velocity $\mathbf{v}$, which identify the mechanical state of the active particles, and over the activity variable u:

$$f_i = f_i(t, \mathbf{x}, \mathbf{v}, u) \; : \; [0, T] \times D_{\mathbf{x}} \times D_{\mathbf{v}} \times D_u \to \mathbb{R}^+, \qquad (6.1)$$

where the subscript $i \in \{1, \ldots, n\}$ refers to each specific subsystem. The function f_i is such that $f_i(t, \mathbf{x}, \mathbf{v}, u) \, d\mathbf{x} \, d\mathbf{v} \, du$ denotes, for the i-th subsystem, the number of active particles whose state, at time t, is in the elementary volume of the space of microscopic states $[\mathbf{x} + d\mathbf{x}] \times [\mathbf{v} + d\mathbf{v}] \times [u + du]$. The whole domain of the microscopic states has been defined by $D_{\mathbf{x}} \times D_{\mathbf{v}} \times D_u \subseteq \mathbb{R}^3$.

Remark 6.1. In general the modeling approach identifies the activity by a scalar variable. The strategy of using a scalar variable, rather than a vector one, is a possible way to reduce analytic complexity. Namely, reducing the dimension of the microscopic state that appears in f_i reduces the structural complexity of the evolu-tion equations for the f_i viewed as dependent variables, although the consequence of a greater number of equations.

The introduction of probability functions to describe the stated of each func-tional subsystem allows computing variables at the macroscopic scale, which are obtained by moments of the generalized distribution function. In particular, the *local density* of the i-th functional subsystem is given by:

$$\rho[f_i](t,\mathbf{x}) = \int_{D_\mathbf{v} \times D_u} f_i(t,\mathbf{x},\mathbf{v},u)\,d\mathbf{v}\,du, \tag{6.2}$$

while, integration over the volume $D_\mathbf{x}$ containing the particles, gives the *total mass* of the i-th subsystem:

$$N[f_i](t) = \int_{D_\mathbf{x}} \rho_i(t,\mathbf{x})\,d\mathbf{x}, \tag{6.3}$$

which depends on time due to the role of proliferative or destructive interactions, as well as to the flux of particles through the boundaries of the volume. The *total mass* $N = N(t)$ of the overall system is given by the sum of all $N_{[\rho_i]}(t)$.

First order moments provide either *linear mechanical macroscopic* quanti-ties, or *linear activity macroscopic* quantities. For instance, the local **activation** and **activation density** are given by:

$$a[f_i](t,\mathbf{x}) = \int_{D_\mathbf{v} \times D_u} u\,f_i(t,\mathbf{x},\mathbf{v},u)\,d\mathbf{v}\,du, \tag{6.4}$$

and

$$A[f_i](t,\mathbf{x}) = \frac{1}{\rho[f_i](t,\mathbf{x})} \int_{D_\mathbf{v} \times D_u} u\,f_i(t,\mathbf{x},\mathbf{v},u)\,d\mathbf{v}\,du. \tag{6.5}$$

Additional moment calculations, for instance second order moments:

$$e[f_i](t,\mathbf{x}) = \int_{D_\mathbf{v} \times D_u} u^2\,f_i(t,\mathbf{x},\mathbf{v},u)\,d\mathbf{v}\,du, \tag{6.6}$$

and

$$E[f_i](t,\mathbf{x}) = \frac{1}{\rho[f_i](t,\mathbf{x})} \int_{D_\mathbf{v} \times D_u} u^2\,f_i(t,\mathbf{x},\mathbf{v},u)\,d\mathbf{v}\,du, \tag{6.7}$$

provide an information on a pseudo-energy that is expressed by the i-th subsystem as a whole.

Remark 6.2. A particular case, namely spatial homogeneity, occurs when the dynamics is space independent and the distribution over the velocity variable is constant in time. In this case the overall state of the system is delivered by the marginal distribution function:

$$f_i = f_i(t,u) \;:\; [0,T] \times D_u \to \mathbb{R}^+, \tag{6.8}$$

where, with abuse of notations, the same letter f_i is still used to avoid an excess of notations. Calculation of the local density, activation, and activation density are obtained as indicated in Eqs. (6.2)-(6.7), where now integration is limited over the variable u only.

Let us now restrict our interest to the *case of spatial homogeneity*, and consider the derivation of class of equations, which describes the evolution of the generalized distribution functions for a system that does not interact with the outer environment (*closed system*).

The evolution system is obtained by the balance of particles in the elementary volume of the microscopic state. This topic is dealt with in various papers [Bellomo, Bianca, and Delitala (2009); Bellomo and Delitala (2008); Bianca (2010)], by means of technically different approaches. Here we provide a unified treatment in view of the specialization at the molecular and cellular scales that will be dealt with in the sequel.

The underlying structure can be written as follows:

$$\partial_t f_i(t,u) = J_i[\mathbf{f}](t,u) = C_i[\mathbf{f}](t,u) + P_i[\mathbf{f}](t,u), \tag{6.9}$$

where $\mathbf{f} = (f_1, \ldots, f_n)$ is the vector of the distribution functions, $J_i = J_i[\mathbf{f}](t,u)$ models the flow, at time t, into the elementary volume $[u, u+du]$ of the space of the microscopic states the i-th functional subsystem due to interactions. The term J_i is split into the sum of two terms $C_i = C_i[\mathbf{f}](t,u)$ and $P_i = P_i[\mathbf{f}](t,u)$ that correspond to two different types of interactions: *conservative interactions*, modeled by the term C_i, for interactions that modify only the microscopic state of particles, and *nonconservative interactions*, modeled by the term P_i, corresponding to interactions that generate proliferation or destruction of particles in their microscopic state.

These interactions are represented in Figures. 6.1, 6.2, and 6.3, corresponding to the visualization, respectively, of long range interactions, destructive, and proliferative encounters.

More in details three types of particles are involved in the interaction process:

• *Test* particles whose distribution function is $f(t,u)$, which interact with field particles and agents;

• *Field* particles whose distribution function is $f(t,u^*)$, which interact with candidate and test particles;

• *Candidate* particles whose distribution function is $f(t,u_*)$, which interact with field particles.

The rules of the interaction dynamics are as follows: *candidate* particles can acquire, in probability, the state of the *test* particles, after an interaction with *field* particles, while test particles lose their state.

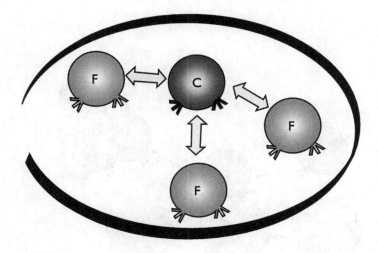

Fig. 6.1 Long range interactions

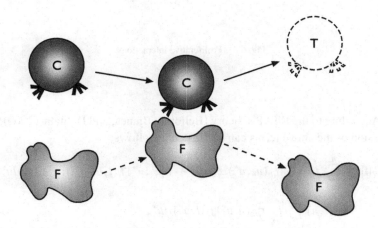

Fig. 6.2 Destructive interactions

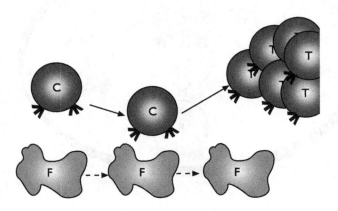

Fig. 6.3   Proliferative interactions

According to the KTAP's theory [Bellomo, Bianca, and Delitala (2009)], the expression of the above terms can be written as follows:

$$C_i[\mathbf{f}] = \sum_{k=1}^{n} \int_{D_u \times D_u} \eta_{ik}(u_*, u^*) \mathscr{B}_{ik}(u_* \to u | u_*, u^*) f_i(t, u_*) f_k(t, u^*) \, du_* \, du^*$$
$$- f_i(t, u) \sum_{k=1}^{n} \int_{D_u} \eta_{ik}(u, u^*) f_k(t, u^*) \, du^*, \tag{6.10}$$

which corresponds to conservative interactions involving active particles, and

$$P_i[\mathbf{f}] = \sum_{h=1}^{n} \sum_{k=1}^{m} \int_{D_u \times D_u} \eta_{hk}(u_*, u^*) \mu_{hk}^i(u_*, u^*; u) f_h(t, u_*) f_k(t, u^*) \, du_* \, du^*$$
$$- f_i(t, u) \sum_{k=1}^{m} \int_{D_u} \eta_{ik}(u, u^*) \rho_{ik}(u, u^*) f_k(t, u^*) \, du^*, \tag{6.11}$$

which corresponds to nonconservative, namely proliferative and/or destructive, interactions.

If the particles of the h (or k) functional subsystem are simply called h-particle (or k-particle), the following descriptions clarify the meaning of the various terms that appear in the above equations:

- η_{ik} is the encounter rate for the encounters of a i-particle (candidate or test), whose state is u_*, with a k-particle (field), whose state is u^*.

- η_{hk} is the encounter rate for the encounters of a h-particle (candidate or test), whose state is u_*, with a k-particle (field), whose state is u^*.

- $\mathscr{B}_{ik}(u_* \to u|u_*,u^*)$ is the probability density that the candidate i-particle with state u_* falls into the state u, of the same i-th functional subsystem, after the interaction with a field k-particle, whose state is u^*.

- $\mu_{hk}^i(u_*,u^*;u)$ models the proliferation into the i-th functional subsystem, due to interactions, which occur with rate η_{hk}, of the candidate h-particle, with state u_*, and the field k-particle, with state u^*.

- $\rho_{ik}(u,u^*)$ models the destruction of particles in the i-th functional subsystem, due to interactions, which occur with rate η_{ik}, of the candidate i-particle, with state u_*, k-particle, with state u^*.

Substituting Eqs. (6.10) and (6.11) into (6.9) yields:

$$
\begin{aligned}
\partial_t f_i(t,u) = J_i[\mathbf{f}](t,u) &= C_i[\mathbf{f}](t,u) + P_i[\mathbf{f}](t,u) \\
&= \sum_{k=1}^n \int_{D_u \times D_u} \eta_{ik}(u_*,u^*)\mathscr{B}_{ik}(u_* \to u|u_*,u^*) f_i(t,u_*) f_k(t,u^*)\,du_*\,du^* \\
&- f_i(t,u) \sum_{k=1}^n \int_{D_u} \eta_{ik}(u,u^*) f_k(t,u^*)\,du^* \\
&+ \sum_{h=1}^n \sum_{k=1}^m \int_D \eta_{hk}(u_*,u^*)\mu_{hk}^i(u_*,u^*;u) f_h(t,u_*) f_k(t,u^*)\,du_*\,du^* \\
&- f_i(t,u) \sum_{k=1}^m \int_{D_u} \eta_{ik}(u,u^*)\rho_{ik}(u,u^*) f_k(t,u^*)\,du^*.
\end{aligned} \tag{6.12}
$$

It is worth stressing the following characteristics of the interaction process that have been taken into account:

i) The strategy developed by the active particles is the output of multiple interactions;

ii) Proliferative events include the generation of active particles into a new functional subsystem as visualized in Figure 6.4. Accordingly mutations generate a number of functional subsystems that evolves in time.

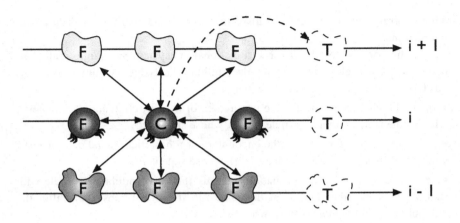

Fig. 6.4   Interactions with transition into a new subsystem

It is convenient, in view of applications, considering also the case of open systems, where each subsystem is subject to m external actions, due to particles, called **agents**, that are modeled as follows:

$$g_{ij} = g_{ij}(t,w) \ : \ [0,T] \times D_w \to \mathbb{R}^+, \qquad (6.13)$$

for $j \in \{1,\ldots,m\}$, which act at the same scale of the active particles for each i-th subsystem. These actions are supposed to be known functions of time and of the variable w. The formal structure of the evolution equations can be written as follows:

$$\partial_t f_i(t,u) = J_i[\mathbf{f}](t,u) + Q_i[\mathbf{f},\mathbf{g}](t,u), \qquad (6.14)$$

where $Q_i = Q_i[\mathbf{f}](t,u)$ refers to interactions between the inner and the outer systems. Such term can be further specialized as follows:

$$Q_i[\mathbf{f}](t,u) = C_i^e[\mathbf{f},\mathbf{g}](t,u) + P_i^e[\mathbf{f},\mathbf{g}](t,u). \qquad (6.15)$$

This splitting shows that two different types of interactions are considered: **conservative interactions** and **nonconservative interactions** between active particles and **field agents**, whose distribution function is denoted by $g_{ik}(t,w^*)$ or $g_{hk}(t,w^*)$.

Calculations analogous to those we have seen previously yield:

$$C_i^e[\mathbf{f}] = \sum_{k=1}^{m} \int_{D_u \times D_w} \eta_{ik}^e(u_*,w^*)\mathscr{C}_{ik}(u_* \to u|u_*,w^*)f_i(t,u_*)g_{ik}(t,w^*)\,du_*\,dw^*$$

$$- f_i(t,u)\sum_{k=1}^{m}\int_{D_w}\eta_{ik}^e(u,w^*)g_{ik}(t,w^*)\,dw^*, \qquad (6.16)$$

and

$$
P_i^e[\mathbf{f}] = \sum_{h=1}^{n} \sum_{k=1}^{m} \int_{D_u \times D_w} \eta_{hk}^e(u_*, w^*) v_{hk}^i(u_*, w^*; u) f_h(t, u_*) g_{hk}(t, w^*) \, du_* \, dw^*
$$
$$
- f_i(t, u) \sum_{k=1}^{m} \int_{D_w} \eta_{ik}^e(u, w^*) \zeta_{ik}(u, w^*) g_{ik}(t, w^*) \, dw^*, \qquad (6.17)
$$

where:

• η_{ik}^e is the encounter rate for the encounters of a i-particle (candidate or test), with state u_*, with a k-agent (field), with state w^*.

• η_{hk}^e is the encounter rate for the encounters of a h-particle (candidate or test), with state u_*, with a k-agent (field), with state w^*.

• $\mathscr{C}_{ik}(u_* \to u | u_*, w^*)$ is the probability density that the candidate i-particle with state u_* falls into the state u after the interaction with a field k-agent with state w^*.

• $v_{hk}^i(u_*, w^*; u)$ models the net proliferation into the i-th functional subsystem, due to interactions, which occur with rate η_{hk}^e, of the candidate h-particle, with state u_*, and k-agent, with state w^*.

• $\zeta_{ik}(u, w^*)$ models the destruction of particles in the i-th functional subsystem, due to interactions, which occur with rate η_{ik}^e, of the candidate i-particle, with state u_*, and k-agent, with state w^*.

Remark 6.3. The distribution functions of the external agents g_{ij} can model both external actions such as therapeutics, or the damage that can derive from the outer environment. It can also model interactions between functional subsystems at different scales.

Remark 6.4. The distribution functions have been assumed to be known functions of their arguments and not modified by interactions. Specific models may include interactions that modify the distribution of specific agents. For instance, this is the case of interactions at different scales. The formal generalization is simply a matter of technical calculations.

The final expression of the evolution equation for the distribution functions of each functional subsystems results to be as follows:

$$\partial_t f_i(t,u) = J_i[\mathbf{f}](t,u) + Q_i[\mathbf{f},\mathbf{g}](t,u)$$

$$= C_i[\mathbf{f}](t,u) + P_i[\mathbf{f}](t,u) + C_i^e[\mathbf{f},\mathbf{g}](t,u) + P_i^e[\mathbf{f},\mathbf{g}](t,u)$$

$$= \sum_{k=1}^{n} \int_{D_u \times D_u} \eta_{ik}(u_*,u^*)\mathscr{B}_{ik}(u_* \to u|u_*,u^*)f_i(t,u_*)f_k(t,u^*)\,du_*\,du^*$$

$$- f_i(t,u) \sum_{k=1}^{n} \int_{D_u} \eta_{ik}(u,u^*)f_k(t,u^*)\,du^*$$

$$+ \sum_{h=1}^{n}\sum_{k=1}^{m} \int_{D_u \times D_u} \eta_{hk}(u_*,u^*)\mu_{hk}^i(u_*,u^*;u)f_h(t,u_*)f_k(t,u^*)\,du_*\,du^*$$

$$- f_i(t,u) \sum_{k=1}^{m} \int_{D_u} \eta_{ik}(u,u^*)\rho_{ik}(u,u^*)f_k(t,u^*)\,du^*$$

$$+ \sum_{k=1}^{m} \int_{D_u \times D_w} \eta_{ik}^e(u_*,w^*)\mathscr{C}_{ik}(u_* \to u|u_*,w^*)f_i(t,u_*)g_{ik}(t,w^*)\,du_*\,dw^*$$

$$- f_i(t,u) \sum_{k=1}^{m} \int_{D_u} \eta_{ij}(u,w^*)g_{ik}(t,w^*)\,dw^*$$

$$+ \sum_{h=1}^{n}\sum_{k=1}^{m} \int_{D_u \times D_w} \eta_{hk}(u_*,w^*)v_{hk}^i(u_*,w^*;u)f_h(t,u_*)g_{hk}(t,w^*)\,du_*\,dw^*$$

$$- f_i(t,u) \sum_{k=1}^{m} \int_{D_w} \eta_{ik}^e(u,w^*)\zeta_{ik}(u,w^*)g_{ik}(t,w^*)\,dw^*. \tag{6.18}$$

Remark 6.5. The mathematical structure (6.18) can be used to model therapeutical actions that are developed at the same scale of the active particles. Specifically, these actions are modeled by the agents with distribution g_{ik} and g_{hk}. Actions at a higher scale, modeled by terms of the type $F_i(t,u)$ act directly on the distribution function according to the following structure:

$$\partial_t f_i(t,u) + \partial_u\Big(F_i(t,u)\,f_i(t,u)\Big) = J_i[\mathbf{f}](t,u), \tag{6.19}$$

for $i \in \{1,\dots,n\}$.

The mathematical structures in the case of space homogeneity can be further developed to take into account space dependent phenomena in several ways as documented by papers [Bellomo, Bianca, and Delitala (2009); Bianca (2010)]. Specifically the first paper technically shows how the same structures can include a dynamics over space and velocity by dealing with an enlarged variable suitable to describe the microscopic state; this variable includes position, velocity, in addition to activity. The contribution of [Bianca (2010)] deals with modeling of different ways to deal with space and velocity variables, regarded as discrete variables, and various generalizations to the case of open systems.

The presentation of this section is limited to an approach that has been specifically used to derive macroscopic tissue equations from the underlying description at the microscopic scale. It basically consists in a stochastic perturbation of the spatially homogeneous dynamics that has been introduced in [Othmer, Dunbar, and Alt (1998)] for particles with homogeneous distribution of the activity variable and used by several authors to derive macroscopic equations [Chalub *et al.* (2004, 2006); Hillen and Othmer (2000)].

The generalization to multicellular systems with heterogeneous distribution of the activity variable has been proposed in paper [Bellomo and Bellouquid (2004)] and further developed towards diffusion limits for mixtures [Bellomo and Bellouquid (2006); Bellomo, Bellouquid, and Herrero (2007)], and hyperbolic limits [Bellomo *et al.* (2007); Bellomo and Bellouquid (2009)].

A generalization of the previous structure (6.18), in the case of space-dependent systems, will be presented in the next chapter. Moreover, the Appendix summarizes the treatment of space structure according to paper [Bellomo, Bianca, and Delitala (2009)]. Accordingly, a stochastic perturbation in velocity of the whole system is considered as follows:

$$\left(\partial_t + \mathbf{v} \cdot \nabla_{\mathbf{x}}\right) f_i(t, \mathbf{x}, \mathbf{v}, u) = \nu_i L_i[f_i] + J_i[\mathbf{f}] + Q_i[\mathbf{f}, \mathbf{g}], \qquad (6.20)$$

where $J_i[\mathbf{f}] = J_i[\mathbf{f}](t, u)$ and $Q_i[\mathbf{f}, \mathbf{g}] = Q_i[\mathbf{f}](t, u)$ are the inner and outer operators that appear in (6.18), ν_i is the turning rate or turning frequency, hence $\tau_i = \frac{1}{\nu_i}$ is the mean run time; the operator $L_i[f_i] = L_i[f_i](t, \mathbf{x}, \mathbf{v})$, which describes the dynamics of biological organisms modeled by a velocity-jump process, has the following form:

$$L_i[f_i] = \int_{D_{\mathbf{v}}} T_i(\mathbf{v}^* \to \mathbf{v}) f_i(t, \mathbf{x}, \mathbf{v}^*, u) \, d\mathbf{v}^* - f_i(t, \mathbf{x}, \mathbf{v}, u) \int_{D_{\mathbf{v}}} T_i(\mathbf{v} \to \mathbf{v}^*) \, d\mathbf{v}^*, \quad (6.21)$$

where $T_i(\mathbf{v}^* \to \mathbf{v})$ is, for the i-th subsystem, the probability kernel for the new velocity $\mathbf{v} \in D_{\mathbf{v}}$ assuming that the previous velocity was $\mathbf{v}^*$.

The above framework is not the most general ones, but simply refers to a stochastic perturbation of the spatially homogeneous case. As we shall see, this structure is sufficient to deal with the objectives of these Lecture Notes. The reader interested to most general cases is addressed to the already cited review paper [Bellomo, Bianca, and Delitala (2009)], which also focuses some applications to the modeling of complex systems. More general frameworks are reported in the Appendix to complete the presentation.

6.3 Guidelines Towards Modeling at the Molecular and Cellular Scales

This section aims at showing how the mathematical structures proposed in the preceding section can be further elaborated to derive models at the molecular and cellular scale. This means treating the following issues:

i) Selection of the functional subsystems that play the game;

ii) Characterization of the activity variables that characterize each subsystem at each scale;

iii) Modeling the general rules that characterize the interaction among active particles.

iv) Derivation of the model by taking advantage of the mathematical structures proposed in the preceding section in agreement with the above outlined strategy.

Usually the derivation of a model requires some simplifying assumptions, hence the solution of the equations is an approximation to the behavior of the original system. The simplified model can be used to give insights into the factors that influence the dynamics of the more complex real system (e.g. the identification of key parameters, or prediction of future behaviors).

The accuracy of the model, in general, depends on the validity of the assumptions and the development of a good model is usually an iterative process, with successive generations of the model suitable to provide better approximations to the real system. Since chemical reactions can affect the behavior of a whole organism (even to the point of causing it to die) any reasonably complex biological or physiological model will include processes that vary on a wide range of time and length scales (multi-scale models).

The mathematical modeling at *cellular scale* can take advantage of the structure defined in Eq. (6.18), where n represents the number of functional subsystems each of them made by a heterogeneous collection of cells that collectively develop the same biological function regarded as the activity variable. Therefore Eq. (6.18) defines the time evolution of the distribution functions f_i over the activity variable for each functional subsystem. Mathematical models are obtained when a mathematical description of the interactions at the cellular scale are introduced to define the various interaction terms $\eta, \mathscr{B}$ and μ. A specific example is delivered in Chapter 8. The structure defined by Eq. (6.20) can be used to derive models at the macroscopic scale as we shall see in Chapter 7.

The above approach can be generalized to deal with the mathematical modeling of phenomena at the molecular scale. The first step consists in the selection of groups of genes whose over and lower expression can cause undesired phenotype expressions at the cellular scale. These groups identify the functional subsystems that play the game.

The representation at the *molecular scale*, namely to the scale of genes grouped into m functional subsystems labeled by the subscript i, is the stochastic expression of proteins for each subsystem. The following distribution function can be used:

$$\varphi_i(t,w) : \quad [0,T] \times \mathbb{R} \to \mathbb{R}^+, \tag{6.22}$$

for $i \in \{1,\ldots,m\}$.

If the average standard expression w_0 is known, the variable w represents the *deviation* from w_0. Therefore, $w > 0$ corresponds to over-expression, while $w < 0$ corresponds to low-expression, and $w = 0$ to the natural expression. Moreover, considering that $dn_i = \varphi_i(t,w)dw$ represents, for the i-subsystem, the density of expression at time t in the interval $[w, w+dw]$, integration over a certain interval $[w_1, w_2]$, namely

$$\int_{w_1}^{w_2} \varphi_i(t,w)\,dw = n_i^{12}(t), \tag{6.23}$$

represents the density of expression at time t in the previously defined interval.

Remark 6.6. The natural expression corresponds to a distribution concentrated over $w = 0$, with density $n_i(t = 0)$, can be used to normalize φ_i. Therefore, the integral of φ_i over $\mathbb{R}$ at $t = 0$ is equal to one. Anomalous expressions are different distributions to be properly identified. The phenotype at cellular level corresponding to anomalous genotype are identified by suitable information delivered by the various distributions φ_i.

Remark 6.7. Modification of the gene expression, namely the dynamics of φ_i is due to interactions between genes of the same or of different functional subsystems as well as to external actions, either therapeutical or other external agents. Specific actions can be applied directly at the molecular scale by actions of the type

$$\psi_k = \psi_k(t,w) : \quad [0,T] \times D_w \to \mathbb{R}^+, \quad k \in \{1,\ldots m\}, \tag{6.24}$$

acting on each population by a given distribution function over microscopic state $w \in D_w$.

The mathematical structure, which describes the evolution of φ_i, is given by Eq. (6.18) that can be simplified by the assumption that interactions do not induce events in a different functional subsystem. Therefore, the following class of equations can be used:

$$\partial_t \varphi_i(t,w) = H_i[\varphi](t,w)$$

$$= \sum_{h=1}^{n} \sum_{k=1}^{m} \int_{D_w \times D_w} \eta_{hk}(w_*, w^*) \mathcal{B}_{hk}(w_* \to w | w_*, w^*) \varphi_h(t, w_*) \varphi_k(t, w^*) \, dw_* \, dw^*$$

$$- \varphi_i(t,w) \sum_{k=1}^{m} \int_{D_w} \eta_{ik}(w, w^*) [1 - \mu_{ik}(w, w^*)] \varphi_k(t, w^*) \, dw^*$$

$$+ \sum_{h=1}^{n} \sum_{k=1}^{m} \int_{D_w \times D_w} \eta_{hk}^e(w_*, w^*) \mathcal{C}_{hk}(w_* \to w | w_*, w^*) \varphi_i(t, w_*) \psi_k(t, w^*) \, du_* \, dw^*$$

$$- \varphi_i(t,w) \sum_{k=1}^{m} \int_{D_w} \eta_{ik}(w, w^*) [1 - v_{ik}(w, w^*)] \psi_k(t, w^*) \, dw^*, \tag{6.25}$$

with obvious meaning of symbols.

If the external action is the same for each gene of a functional subsystem, then the structure is as follows:

$$\partial_t \varphi_i(t,w) + F_i(t) \partial_w \varphi_i(t,w) = K_i[\varphi](t,w)$$

$$= \sum_{h=1}^{n} \sum_{k=1}^{m} \int_{D_w \times D_w} \eta_{hk}(w_*, w^*) \mathcal{B}_{hk}(w_* \to w | w_*, w^*) \varphi_h(t, w_*) \varphi_k(t, w^*) \, dw_* \, dw^*$$

$$- \varphi_i(t,w) \sum_{k=1}^{m} \int_{D_w} \eta_{ik}(w, w^*) [1 - \mu_{ik}(w, w^*)] \varphi_k(t, w^*) \, dw^*, \tag{6.26}$$

where $F_i(t)$ models the external action over the i-th functional subsystem, which in the nonlinear case writes as follows:

$$\partial_t \varphi_i(t,w) + \partial_w \Big(F_i(t,u) \, \varphi_i(t,w) \Big) = K_i[\mathbf{f}](t,u), \tag{6.27}$$

for $i \in \{1, \dots, m\}$.

6.4   Additional Analysis Looking at the Immune Competition

Modeling the immune competition is one of the most challenging research field for applied mathematicians involved in the development of mathematical tools to treat complex systems. Indeed, the immune system shows all complexity characteristics related, among various ones, to the large number of different types of entities involved in the game, the heterogeneous expression of the defence strategy, and the complex hiding and learning dynamics, for short H-L-dynamics, developed during the competition.

As already mentioned in Chapter 3, the immune system plays an important role in the dynamics of the biological phenomena under consideration. In fact, it can activate a defence of immune cells against infectious agents and, in general, against cells that are carriers of pathological states. The related dynamics also involves cells that, due to inflammation, may progressively degenerate into cancer cells. This topic has been technically reported in Chapter 4 based on [Hanahan and Weinberg (2000)], where all conceivable mutations that lead to the onset of cancer cells, have been briefly described.

This section proposes some preliminary ideas to analyze how KTAP's methods and the system theory approach can be properly developed towards the modeling of the complex dynamics of the immune competition, which involves both mutations, hiding, and learning processes.

The immune competition is generated whenever a pathological state is present. Specifically, it acts in all different phases of the onset of wounds and healing processes described in Part I. The approach of this section is essentially methodological, however it will be technically applied to the next one where the modeling of the immune defence is included in the analysis of keloid formation and degeneration that, if not sufficiently contrasted, may even end up with the onset of cancer cells.

An important difficulty that needs to be treated is generated by the presence of cells, either host or guest, that mutate progressively. These mutations produce a number of new species that grows in time and put in difficulty the learning process of the immune system. Generally, mutations occur during cell reproduction simply by errors in the DNA replication. However, it may happen that cells with a higher adaptive ability survive and reproduce. For instance, this is the case of cells with proliferative ability that is not followed by apoptosis. These cells are initially useful to repair processes but may subsequently undergo further mutations acquiring the competence of cancer cells. Accordingly, the whole process follows rules of a Darwinian type evolution, where progressive mutations may even occur in short time lapses. A phenomenological description is delivered by Chapter 4. As a paradox, the immune system when it is not able to suppress the undesired presence of abnormal carriers ultimately contributes to their progression by selecting stronger carriers and suppressing the weaker ones.

The state-of-the-art indicates that mathematics, in spite of several interesting contribution has not yet provided a fully satisfactory approach to the highly difficult problem under consideration. The same indication can be stated from the viewpoint of biological sciences. The interested reader can find the pertinent literature concerning the immune competition from the viewpoint of immunologists by the review papers [Cooper (2010); Goldstein, Faeder, and Hlavaceck (2004)] and therein cited bibliography.

The contents refer to the mathematical framework reported in the preceding sections. Further reference, is given by the phenomenological descriptions and by the general strategies to reduce complexity that have been proposed in Part I. The approach technically refers to the content of [Bellomo (2010)], that takes advantage of some perspective ideas proposed in [Cattani and Ciancio (2007, 2008)], where the authors derive a model of population dynamics whose coefficients are related to the dynamics at the cellular scale.

The presentation of mathematical tools is focused, with tutorial aims, on a minimal model that at $t = 0$ consists of two functional subsystems only, namely cells that are carrier of a pathological state and immune cells. Specifically, we refer to the following structure:

$$\partial_t f_i(t,u) = J_i[f_i, f_j](t,u) = \sum_{j=1}^{2} J_{ij}[\mathbf{f}](t,u)$$

$$= \sum_{j=1}^{2} \int_{D_u \times D_u} \eta_{ij}(u_*, u^*) \mathcal{B}_{ij}(u_* \rightarrow u | u_*, u^*) f_i(t, u_*) f_j(t, u^*) du_* du^*$$

$$- f_i(t,u) \int_{D_u} \eta_{ij}(u, u^*)[1 - \mu_{ij}(u, u^*)] f_j(t, u^*) du^*, \qquad (6.28)$$

where all terms and the interaction dynamics have been defined in Section 6.2.

The rules of the interaction dynamics are as follows: *candidate* particles can acquire, in probability, the state of the *test* particles, after an interaction with *field* particles, while test particles lose their state. Interactions, modeled by the terms $\mathcal{B}_{ij}$, have been called *stochastic games* considering that active particles whose state is known in probability, and that the output is identified by a probability density are involved. Moreover, interactions of *test* with *field* particles may generate proliferation or destruction of the text particles.

This structure has been applied by various authors [Arlotti, Gamba, and Lachowicz (2002); De Angelis and Jabin (2003, 2005); Kolev (2003); Kolev2 (2003); Kolev, Kozlowska, and Lachowicz (2005)] and many others, as documented in the review paper [Bellomo and Delitala (2008)], to model the competition between cancer and immune cells. Here a development of such structure is proposed as follows.

An important concept that is useful to the analysis developed in what follows, is the definition of a **distance** α_{ij} between the active particles of the i-th and the j-th functional subsystems. The following hierarchy, that corresponds to increasing level of sophistication, is proposed:

i) $\alpha_{ij} = \eta_{ij}^0 \in \mathbb{R}$ is simply a real constant that depends on the interacting functional subsystems.

ii) $\alpha_{ij} = \alpha_{ij}(u_{*i}, u_j^*)$ depends on the difference among the microscopic states of the interacting particles:

$$\alpha_{ij}(u_{*_i}, u_j^*) = |u_{*_i} - u_j^*|.$$

iii) $\alpha_{ij} = \alpha_{ij}[f_i, f_j](t)$ depends on the distance among the overall state of the interacting functional subsystems:

$$\alpha_{ij}[f_i, f_j](t) = \sqrt{\int_{D_u} [f_i - f_j]^2(t, u) \, du}. \tag{6.29}$$

Of course additional examples can be selected according to the specific characteristics of the system under consideration. However, the analysis is restricted, with tutorial aims, to the three simple examples given above, where the difference is that the distance, respectively, is a constant, depends on the state of the interacting pairs, and is a functional of the probability distributions that characterize the two interacting functional subsystems.

The modeling of encounter rate η_{ij} can be achieved in a fashion that increasing values of the distance α_{ij} correspond to decreasing values of the encounter rate η_{ij}. According to the previous hierarchy we define the encounter rate as follows:

i) The encounter rate is a constant simply given by

$$\eta_{ij} = \eta_{ij}^0. \tag{6.30}$$

ii) The encounter rate is related to the difference among the microscopic state of the interacting particles as follows:

$$\eta_{ij} = \eta_{ij}(u_{*_i}, u_j^*) = \eta_{ij}^0 \, e^{-c\,\alpha_{ij}^2(u_{*_i}, u_j^*)}, \tag{6.31}$$

where c is a positive real constant.

iii) The encounter rate is related to the distribution functions of the interacting functional subsystems:

$$\eta_{ij} = \eta_{ij}(t|f_i, f_j) = \eta_{ij}^0 \, e^{-c\,\alpha_{ij}^2[f_i, f_j](t)} = \eta_{ij}^0 \, e^{-c\left(\int_{D_u} [f_i - f_j]^2(t, u)\, du\right)}. \tag{6.32}$$

Therefore equation (6.28) is rewritten as follows:

$$\partial_t f_i(t, u) = J_i[f_i, f_j](t, u) = \sum_{j=1}^{2} J_{ij}[\mathbf{f}](t, u)$$

$$= \sum_{j=1}^{2} \int_{D_u \times D_u} \eta_{ij}(u_*, u^*|f_i, f_j) \, \mathcal{B}_{ij}(u_* \to u|u_*, u^*) \, f_i(t, u_*) \, f_j(t, u^*) \, du_* \, du^*$$

$$- f_i(t, u) \int_{D_u} \eta_{ij}(u, u^*|f_i, f_j) [1 - \mu_{ij}(u, u^*)] \, f_j(t, u^*) \, du^*, \tag{6.33}$$

where the formal difference with respect to Eq. (6.28) is that now the encounter rate η_{ij} can be conditioned by the distribution functions of the interacting active particles. Therefore, additional developments are necessary to characterize such a difference otherwise simply formal.

Remark 6.6. The distance (6.29) between the two functional subsystems evolves in time according to the following equation:

$$\partial_t \alpha_{ij} = \frac{1}{\sqrt{\int_{D_u} (f_i - f_j)^2 (t, u) \, du}} \int_{D_u} \left(J_i[\mathbf{f}](t, u) - J_j[\mathbf{f}](t, u) \right) du. \tag{6.34}$$

Focusing on proliferative and/or destructive events, we have seen in Section 6.2 that this dynamics includes the onset of new functional subsystems with characteristics that reduce the learning ability of the hunting subsystem. In this case, it is useful to introduce a further modification of the mathematical structure (6.28) as follows:

$$\partial_t f_i(t, u) = G_i[\mathbf{f}](t, u) = \sum_{h=1}^{p} \sum_{k=1}^{p} G_{hk}[\mathbf{f}](t, u)$$

$$= \sum_{h=1}^{p} \sum_{k=1}^{p} \int_{D_u \times D_u} \eta_{hk}^0 \, e^{-c \, \alpha_{hk}^2 (t | f_h, f_k)} \, \mathscr{B}_{hk}^i (u_* \to u | u_*, u^*) \, f_h(t, u_*) \, f_k(t, u^*) \, du_* \, du^*$$

$$- \sum_{h=1}^{p} \sum_{k=1}^{p} \int_{D_u} \eta_{hk}^0 \, e^{-c \, \alpha_{ik}^2 (t | f_h, f_k)} [1 - \mu_{hk}^i (u_*, u^*)] f_h(t, u_*) \, f_k(t, u^*) \, du_* \, du^*, \tag{6.35}$$

where:

- $\mathscr{B}_{hk}^i$ is the probability density that a candidate particle of the h-th functional subsystem, and with state u_*, ends up in the state u of the i-th functional subsystem after the interaction with the field particle, with state u^*, of the k-th functional subsystem. This term satisfies, for all $h, k \in \{1, 2\}$, the following condition:

$$\sum_{i=1}^{p} \int_{D_u} \mathscr{B}_{ij} (u_* \to u | u_*, u^*) \, du = 1, \quad \forall u_*, u^* \in D_u.$$

- μ_{hk}^i is the proliferative/destructive particle of the h-th functional subsystem, with state u_*, into the state u of the i-th functional subsystem due to the encounter with the particle (field) of the k-th functional subsystem, with state u^*.

The mathematical structure (6.35) involves a variable number of equations, namely $p = 2$ at $t = 0$, with $p = p(t)$ due to the onset of new functional subsystems due both to conservative and/or destructive events, modeled, respectively, by the terms $\mathscr{B}$ and/or μ.

Equation (6.33) can act as a paradigm to model the immune competition. Some guidelines are here proposed to pursue such objective in view of a more detailed analysis that will be developed in the sequel:

6.1. The complexity of the overall system due to the large number of interacting entities can be reduced by decomposing it into two functional subsystems labeled by the subscript $i \in \{1,2\}$; where $i = 1$ corresponds to cells that are carrier of a pathological state, and $i = 2$ to immune cells. This decomposition is valid at $t = 0$, while for increasing time the number functional subsystems may increase due to mutations of the cells of the first subsystem.

6.2. The activity variable u models the progression of the pathological state for $i = 1$ and defence ability for $i = 2$. For both subsystems $u \in \mathbb{R}$, where positive values identify the afore-mentioned ability, while negative values denote the opposite behavior.

6.3. The modeling approach should define a **_distance_** α_{ij} between the i-th and j-th functional subsystems. Subsequently, the encounter rate η_{ij} between them is referred to such a distance, while the H-L-dynamics is described by the transition density $\mathscr{B}_{ij}$. The encounter rate decreases with increasing distance.

6.4. Interactions generate proliferative and/or destructive events modeled by the terms μ_{ij} corresponding to encounters that occur with rate η_{ij}. Proliferation can generate new functional subsystems that are related to mutations so that the number of functional subsystems that play the game changes in time.

These guidelines have to be transferred in mathematical equations suitable to develop the modeling approach to the H-L dynamics such that cells of the first population modify their shape and consequently hide their presence to immune cells. These, conversely, learn the presence of cells of the first functional subsystem and develop their defence strategy.

6.5 Critical Analysis

This chapter has presented some mathematical structures that can be used to derive specific models corresponding to the biological system described in Part I. The conceptual approach can be referred to that used in cancer modeling [Bellomo and Delitala (2008); Bellomo, Li, and Maini (2008); Jackiewicz (2009); Drucis *et al.* (2010)], which is characterized by some analogy with the specific pathologies presently under consideration. In fact, the system evolves in time, while proliferative and destructive events take place. The organism reacts to the presence of a pathologic state, for instance in the case of the wound healing processes, by a competitive process which may end up with the repair or with a regression of the pathology.

The scientific community agrees that the various pathologies under consideration can be regarded as genetic disease, as many other diseases, and that the struggle with the organism is contrasted by the immune system and possibly by specific therapeutical actions possibly directed to cells and genes. Indeed, an analogous dynamics is witnessed in the healing and repair of biological tissues with the exception that healing may be caused, also but not only, by an external action. However, the subsequent dynamics involves all conceivable scales from the low to the higher scale. Namely the dynamics at the molecular scale, that may be influenced by the interactions with the outer environment, determines the dynamics at the cellular scale. The structure, hence the mathematical models, of tissues depends on the underlying description at the cellular scale.

The mathematical structures proposed in this chapter constitute the background for the derivation of models according to a mathematical approach based on the concept that the intensity of the biological functions are heterogeneously distributed among the entities that compose the system and express these functions. Moreover, the system evolves in time due to interaction among the entities themselves and with the outer environment. In some cases, the evolution may include mutations related to dynamics at the molecular scale.

An additional issue that it is worth to be mentioned is the role of nonlinear interactions that has still to be regarded as a challenging open problem waiting for a self-consistent mathematical approach. Some perspective ideas can be given, referring to [Bellomo (2010)], that contribute also to a deeper understanding of the applications proposed in Part III.

Some reasonings can be referred to one of the various mathematical structures presented in Section 6.2, for instance to Eq. (6.12), where interactions related to the terms η, $\mathscr{B}$ and μ are linear, namely by superposition of the actions of field particles to the candidate or test particles. Moreover, it has been assumed that the encounter rates and the output of the interaction depend on the microscopic state, namely the activity variables, of the interacting pairs, while a more general treatment should refer them to the distribution function of the interacting active particles. A useful contribution to the modeling approach is offered by the mathematical description of the hiding-learning dynamics that has been presented in the preceding section. However, this approach is still at a preliminary stage to be properly developed within a proper research program.

In general, the above scenario can be transferred into a consistent mathematical framework, hopefully a mathematical theory, only if the links joining the various scales is constructed. The present state-of-the-art is only a partial contribution to models at each scale, while the afore-mentioned links have not yet been deeply analyzed.

Finally, let us comment that all previous equations have been derived under the assumption of linear, although multiple, interactions and by supposing that the encounter rate depends on the state of the interacting pair. Recent activity in the field suggests to consider nonlinear interactions and encounter rates that depend on the distribution function of the interacting active particles. These developments lead to equations with higher order of nonlinearities as it will be discussed in the last section of this chapter and in the Appendix, where suitable bibliographical indications will be given.

Of course, the authors do not naively claim uniqueness of the mathematical approach presented in this monograph. The interested reader is addressed to the review [Auger *et al.* (2008)] for models at the super-macroscopic scale where the state of the system is represented by the number of individuals in different populations. Population models with internal structure can be modeled by the approach initiated in [Webb (1985)], namely by deterministic partial differential equations. This method has been subsequently developed by various authors as documented by the book [Dieckmann and Heesterbeck (2000)] devoted both to mathematical foundations and applications to biology. The derivation of evolution equations for the probability distribution, charged to describe in the model, the probability distribution over the microscopic state of the system has been proposed in [Schweitzer (2003)]. All these different approaches can be successful to model some of the various biological phenomena presented in this monograph.

Chapter 7

Multiscale Modeling: Linking Molecular, Cellular, and Tissues Scales

7.1 Introduction

The preceding chapter has presented, in the spatially homogeneous case, the mathematical tools of the kinetic theory for active particles which offer a mathematical framework for the derivation of models at the molecular and cellular scales. On the other hand, a space structure is needed to model cellular motion, as well as to recover macroscopic models from the underlying microscopic description. However, the biological phenomena under consideration need a multiscale approach to link all the scales, namely molecular, cellular, and tissue, through the time evolution of the system.

This chapter deals with the mathematical approach to model the links with the lower and higher scales of genes and tissues respectively. The analysis includes the role of the immune system that may contribute to contrast pathological states, but may also negatively contribute to a Darwinian-type selection of cells carriers of aggressive undesired states. Moreover, the derivation of tissue equations leads models generated by the underlying description at the cellular scale. The structure of these equations evolves in time due to the natural progression of cells that is generally accelerated in the case of pathological states.

The derivation of macroscopic (tissue) equations from models at the cellular scale appears to be a complex problem, due to the evolution in time of both the biological functions and of the number of cells. The ratio between the various rates characterizing each evolution - biological, mechanical, and proliferative/destructive - plays an important role in assessing the structure of mathematical macroscopic equations derived from the underlying microscopic equations.

The structure of biological tissue equations evolves in time due to the natural progression of cells, which is generally accelerated in the case of pathological states. This approach is in contrast with the traditional one that *a priori* postulates the structure of tissue models. Time and space evolutions play an important role in determining the dynamics of biological systems. Therefore, we believe that the contents of this chapter should contribute to a mathematical biological theory, as will be critically analyzed in the last chapter of this monograph.

Details on the contents of this chapter can now be given after the afore-mentioned reasonings. Section 7.2 presents a brief survey of the mathematical models obtained from the purely phenomenological approach at the macroscopic scale. These models are obtained by means of classical reaction diffusion and/or conservation equations closed by phenomenological models of the material be-havior of the tissue. However, we should mention that we conceptually support the idea that mathematics should allow the derivation of models at the higher scale from the underlying description at the lower scale, namely tissue models should be derived from models at the cells scale, the dynamics of which is ruled by the molecular scale. This approach differs from the purely phenomenological deriva-tion, considering that the properties of living tissues evolve in time, as documented in Part I. Therefore, this chapter is simply a reference for the derivation from the underlying description at the cells scale which may be obtained using the methods proposed hereafter.

Sections 7.3 deals with the mathematical methods to derive macroscopic equations from the underlying description offered at the cellular level by means of the mathematical structures we saw in Chapter 6. The analysis is developed for a closed system that does not interact with the outer environment, first for a sin-gle subsystem, and subsequently for binary mixtures that offer a more interesting framework for applications in biology, where several components generally play the game.

Section 7.4 shows how the mathematical approach can be extended to several interacting subsystems in the case of open systems where interactions with the outer environment are taken into account.

Section 7.5 treats the challenging problem of linking cellular dynamics to the underlying dynamics at the molecular scale of genes. The approach reviewed in this section is limited to perspective ideas and mathematical tools in view of the conceivable applications.

Finally, Section 7.6 proposes a critical analysis that is mainly focused on the analytic problems generated by the contents of this chapter.

7.2 On the Phenomenological Derivation of Macroscopic Tissue Models

The traditional method to derive tissue level equations is based on the classical approach of continuum mechanics that uses mass and momentum conservation equations properly closed by phenomenological models corresponding to the material behavior of the system. Some review papers, such as [Bellomo, De Angelis, and Preziosi (2003)] and [Bellomo, Li, and Maini (2008)], provide the conceptual background and show how different models are obtained according to the different ways chosen to close conservation and equilibrium equations.

More recent research activity, documented in the book [Chauviere, Preziosi, and Verdier (2010)], shows how the theoretical and experimental investigation focused on the material behavior of biological tissues can contribute to improve the material models needed for the closure. On the other hand, the information delivered by experiments is generally valid in equilibrium conditions, while the system under consideration operates far from equilibrium and is subject, as we have seen, to mutations, which amounts to time evolution of the material properties.

Mathematical modeling of biological complex phenomena, such as fibrosis diseases, should be based on a multiscale approach that accounts for phenomena occurring at different temporal and spatial scales. Unfortunately, most of the models in the literature are macroscopic, either continuum or discrete. As already mentioned, continuum models are derived by the classical approach of continuum mechanics namely by writing the conservation equations for mass and linear momentum and by closing them using phenomenological models corresponding to the material behavior of the system. Generally, mechanical effects are ignored and a specific assumption for the movement is made. However, it is still useful presenting a brief survey not only for the sake of completeness, but also to describe mathematical structures and models that can be subsequently compared with those obtained by the asymptotic analysis presented in this chapter.

A different way for closing the system is to use mechanical models where stress and strain responsible for cellular and tissue deformation are taken into account. In these mechanistic models an extra equation for force balance is used to determine how cells and tissue move as a result of forces action.

Macroscopic models are based on the interactions between cells and extracellular matrix, as cells may synthesize and undergo degradation, which may affect cell properties and orientation (in dermal wound healing this inter-dependent relationship is called *dynamic reciprocity*, see [Clark (1993)]). The evolution of both cells and extracellular matrix is mediated by some chemical growth factors, which regulate both cell proliferation and extracellular matrix reorganization.

The biological quantities appearing in continuum models for wound healing and fibro-proliferative diseases obey the generic conservation equation

$$\partial_t w_i(t, \mathbf{x}) = -\nabla_{\mathbf{x}} \cdot \mathbf{J}_i(t, \mathbf{x}, \mathbf{w}) + S_i(t, \mathbf{x}, \mathbf{w}), \tag{7.1}$$

where w_i is the quantity of the material of the functional subsystem, $\mathbf{J}_i$ represents the flux of w_i and S_i the production/degradation rate. These terms are modeled by phenomenological interpretation of the biological reality.

A very simple model, see [Olsen, Sherratt, and Maini (1996)], is as follows:

$$\begin{cases} \partial_t n = \partial_x \left[D_n(n) \, \partial_x n - f(n) \, \partial_x c \right] + S_n(n, c), \\[2ex] \partial_t c = \partial_x \left[D_c(c) \, \partial_x c \right] + S_c(n, c), \end{cases} \tag{7.2}$$

where $n(x,t)$ is the fibroblast density (the only cell species in the model) and $c(x,t)$ is the growth factor, and S_n and S_c model the source terms. See also [Olsen, Sherratt, Maini (1995); Olsen, Sherratt, and Maini (1996); Olsen *et al.* (1998, 1999)].

This model refers to the following more general structure:

$$\begin{cases} \partial_t n_i = \nabla_{\mathbf{x}} \cdot \left(D_i \nabla_{\mathbf{x}} n_i - \chi(n, c) \nabla_{\mathbf{x}} c - n_i a_i \nabla_{\mathbf{x}} \rho - n_i \partial_t \mathbf{u} \right) + \Gamma_i(n, c), \\[2ex] \partial_t c = \nabla_{\mathbf{x}} \cdot \left(D_c \nabla_{\mathbf{x}} c - c \partial_t \mathbf{u} \right) + \Sigma(n, c), \\[2ex] \partial_t \rho = \nabla_{\mathbf{x}} \cdot \left(-\rho \partial_t \mathbf{u} \right) + \Phi(n, c, \rho), \\[2ex] \nabla_{\mathbf{x}} \cdot \sigma = \rho \, \Omega(\mathbf{u}), \end{cases} \tag{7.3}$$

where:

$n_i(\mathbf{x}, t)$ represents the cell density of the i-th species of cells involved in the model $(n = (n_1, \ldots, n_m))$;

D_i is the diffusion coefficient of the i-th species, usually assumed to be constant;

$c(\mathbf{x}, t)$ is the chemical (growth-factor). The usual hypothesis is to consider it as a generic growth factor, in other words a generic chemical tracking the average behavior of chemicals that take part in the system;

$\chi(n, c)$ is the chemotactic sensitivity factor;

a_i is the haptotactic constant of the i-th species;

$\rho(\mathbf{x}, t)$ is the collagen, or extracellular matrix, density . As for the chemical ρ possesses the average characteristic of various type of collagens forming the extracellular matrix;

$\mathbf{u}(\mathbf{x}, t)$ is the tissue displacement;

σ is the stress tensor;

Γ_i, Σ, Φ and Ω are kinetic terms.

Still referring to [Fusi (2009)] the specific meaning of each equation is as follows:

• Equation $(7.3)_{(1)}$ represents mass balance of the i-th species. The contribution to the flux is given by linear Fickian diffusion, growth factor mediated chemotaxis (where $\chi(n, c)$ is suitably chosen according to the cell surface receptor mechanism), linear haptotaxis and passive convection due to tissue displacement. The kinetic term $\Gamma_i(n, c)$ may include mitosis, phenotypic transformation between different cell species and natural cell death (apoptosis).

• Equation $(7.3)_{(2)}$ represents mass balance of the growth factor c. The flux is driven by linear diffusion and convection, while the term $\Sigma(n, c)$ accounts for production and consumption of chemicals by cells and natural chemical decay.

• Equation $(7.3)_{(3)}$ represents mass balance of the extracellular matrix. The flux is assumed to be due only to convection since collagen fibers are generally linked in a mesh-like structure so that the effects of diffusion is negligible (see [Clark (1989)], [Jennings and Hunt (1992)]). The kinetic term $\Phi(n, c, \rho)$ includes synthesis of collagen by cells, enhancement due to growth-factor chemical and degradation.

• Equation $(7.3)_{(4)}$ represents tissue linear momentum balance, where σ is the stress tensor and $\Omega(\rho, \mathbf{u})$ accounts for external forces acting on the tissue. System (7.3) is a mechanistic model. The stress σ consists of the contribution from the extra cellular matrix and the cells so that: $\sigma = \sigma_{ECM} + \sigma_{cell}$.

Usually, under the assumption of isotropy for ECM fibers alignment and cell orientation, a simple visco-elastic constitutive relation, with stress/strain linear dependence see [Landau, Lifshitz (1970)], is used for σ_{ECM}. The stress exerted by the cells σ_{cell}, on the other hand, has to take into account the enhancement of cell traction force due to high cell density and the inhibition due to high collagen density [Murray, Maini, and Tranquillo (1988)].

Of course, as some authors have remarked [McCarthy, Sas, and Furcht (1988); Murray and Oster (1984); Stopak and Harris (1982)], the hypothesis of isotropy for the extracellular matrix in the presence of cell force traction is definitely strong. For this reason some models take into account anisotropy. In particular, two alignment mechanisms have been proposed [Olsen *et al.* (1999)]: flux-induced alignment and stress-induced alignment. The first is caused by the remodeling of the extracellular matrix as cells move through the tissue, while the second results from mechanical forces acting in the wound site. More in general, this type of modeling can be improved by recent developments of biological mixture theory [Mollica, Preziosi, and Rajagopal (2007)] based on foundation of mixtures including mixtures for growing tissues [Hunprey and Rajagopal (2002)].

The interested reader can find in the already cited paper [Fusi (2009)] various additional details that are not reported here also considering that we are actually interested in the derivation of macroscopic equations from the underlying microscopic description as shown by the methods presented in the next sections.

Therefore, an alternative approach consists in deriving the macroscopic behavior from the dynamics at the cellular level, including mutations. Namely, macroscopic models should be derived from the underlying cellular models by letting intercellular distances tend to those of the tissue level. Possibly, the molecular dynamics should be related to that at the molecular scale. Moreover, various space phenomena, including invasion and pattern formation due to aggregation and chemotaxis [Chalub *et al.* (2004, 2006); Dolak and Schmeiser (2005); Erban and Othmer (2004); Filbet, Laurençot, and Perthame (2005)], play a relevant role in the overall dynamics.

The above approach is widely studied in the case of classical particles by asymptotic methods developed by the mathematical kinetic theory. In recent years, the analysis of the applicability of asymptotic methods has reached an important development according to both the so-called parabolic and hyperbolic limits, or equivalently low and high field limits. The parabolic (low field) limit of kinetic equations leads to a drift–diffusion type system (or reaction–diffusion system) in which the diffusion processes dominate the behavior of the solutions. The review by Bonilla and Soler [Bonilla and Soler (2001)] provides an exhaustive description of the technical differences that characterize the two technically different approaches.

In principle, the same methodological approach can be developed based on multicellular models obtained by methods of the kinetic theory for active particles, however various additional difficulties have to be considered. For instance the generation of proliferative and/or destructive phenomena, mutations that lead to a variable number of equations and the role of external therapeutical actions that possibly contrast the degeneration of tissues.

7.3   Cellular-Tissue Scale Modeling of Closed Systems

The review proposed in this section is focused on the derivation of macroscopic equations in the simple case of one and two functional subsystems, closed with respect to external actions. The next section shows how the approach can be developed in the case of variable number of functional subsystems. In fact, mutations generate new functional subsystems that end up with additional equations of the overall system viewed as a mixture. The analysis is focused on the hyperbolic limit according to the authors' opinion that propagation occurs with finite velocity. However, the bibliography on the diffusive limits is given to provide a complete presentation of the subject.

The mathematical structure to be used toward the analysis that has been outlined above is given by Eqs. (5.19) and (5.20), which is a technical development of the approach initiated by Othmer *et al.*, [Hillen and Othmer (2000); Othmer, Dunbar, and Alt (1998); Othmer and T. Hillen (2002)], obtained by adding to the space homogeneous description stochastic velocity jump process. Subsequently various authors have developed the modeling method including additional biological phenomena such as proliferative/destructive interactions related to the dynamics at the cellular level, see [Bellomo and Bellouquid (2004, 2006); Bellomo, Bellouquid, and Herrero (2007); Lachowicz (2005)]. In these papers biological systems are considered for which interactions do not follow classical mechanical rules, and biological activity may play a relevant role in determining the dynamics. On the other hand, the analysis developed in [Bellomo and Bellouquid (2009)] for a system constituted by one functional subsystem only, and in [Bellomo, Bellouquid, and Herrero (2007)] for a system of two interacting functional subsystems (limited, however, to the case of mass conservative encounters), shows that interactions that change the biological functions of cells may substantially modify the structure of the macroscopic equations. In particular, the analysis proposed in [Bellomo and Bellouquid (2006)] has shown the onset of linear and nonlinear diffusion terms departing from the simple mass conservation equation. A particularly interesting result is the appearance of source terms in the case of proliferative phenomena for systems consisting in, at least, two populations.

7.3.1 *Asymptotic Methods for a Single Subsystem*

This section deals with the derivation of macroscopic equations for a biological system where the space variable has a relevant meaning. The system is constituted by a large number of heterogeneous interacting particles whose microscopic state is characterized by three variables $\{\mathbf{x}, \mathbf{v}, u\}$, where $(\mathbf{x}, \mathbf{v}) \in D_{\mathbf{x}} \times D_{\mathbf{v}} \subseteq \mathbb{R}^3 \times \mathbb{R}^3$, is the *mechanical microscopic state*, related to position and velocity of the particles, and the scalar variable $u \in D_u \subseteq \mathbb{R}$ is the *biological microscopic state*, related to the biological function expressed by collected particles regarded as an only functional subsystem.

The statistical collective description of the overall state of the functional subsystem is defined by the generalized distribution function

$$f = f(t, \mathbf{x}, \mathbf{v}, u) \; : \; [0, T] \times D_\mathbf{x} \times D_\mathbf{v} \times D_u \to \mathbb{R}^+, \text{over position} \, \mathbf{x},$$

velocity $\mathbf{v}$, and the activity variable u. The function f is such that $f(t, \mathbf{x}, \mathbf{v}, u) \, d\mathbf{x} \, d\mathbf{v} \, du$ denotes, under suitable local integrability conditions, the number of active particles whose state, at time t, is in the elementary volume

$$[\mathbf{x}, \mathbf{x} + d\mathbf{x}] \times [\mathbf{v}, \mathbf{v} + d\mathbf{x}] \times [u, u + du]$$

of the space of microscopic states. As already reported in Chapter 6, when this function is known, then macroscopic gross variables can be computed, under suitable integrability properties, by weighted moments.

Following the approach proposed in [Bellomo *et al.* (2007, 2010)], we assume that the transport in position is linear with respect to the velocity and consider a stochastic perturbation in velocity. Thus the time evolution of f is represented by the following equation:

$$(\partial_t + \mathbf{v} \cdot \nabla_\mathbf{x}) \, f(t, \mathbf{x}, \mathbf{v}, u) = \nu L[f](t, \mathbf{x}, \mathbf{v}) + C[f](t, \mathbf{x}, \mathbf{v}, u) + P[f](t, \mathbf{x}, \mathbf{v}, u), \quad (7.4)$$

where:

• ν is the turning rate or turning frequency of the velocity-jump, hence $\tau = \frac{1}{\nu}$ is the mean run time;

• The operator $L[f] = L[f](t, \mathbf{x}, \mathbf{v})$, which has been proposed by various authors to model the dynamics of biological organisms by a velocity-jump process, is defined as follows:

$$L[f] = \int_{D_\mathbf{v}} T(\mathbf{v}^*, \mathbf{v}) f(t, \mathbf{x}, \mathbf{v}^*, u) \, d\mathbf{v}^* - f(t, \mathbf{x}, \mathbf{v}, u) \int_{D_\mathbf{v}} T(\mathbf{v}, \mathbf{v}^*) \, d\mathbf{v}^*, \quad (7.5)$$

where $T(\mathbf{v}, \mathbf{v}^*)$ is the probability kernel over the new velocity $\mathbf{v} \in D_\mathbf{v}$, assuming that the previous velocity was $\mathbf{v}^*$. Moreover, the set $D_\mathbf{v}$ of possible velocities is assumed bounded and spherically symmetric, i.e. $\mathbf{v} \in D_\mathbf{v} \Rightarrow -\mathbf{v} \in D_\mathbf{v}$. Accordingly, particles of the functional subsystem are able to choose any direction with bounded velocity. It is worth mentioning that, in more general cases that are not considered in this monograph, the operator T may depend on f;

• The nonlinear operators $C[f] = C[f](t, \mathbf{x}, \mathbf{v}, u)$ and $P[f] = P[f](t, \mathbf{x}, \mathbf{v}, u)$, corresponding, respectively, to conservative and proliferative/destructive interactions, have the following forms:

$$C[f] = \int_{D_u \times D_u} \eta \mathscr{B}(u_* \to u | u_*, u^*) f(t, \mathbf{x}, \mathbf{v}, u_*) f(t, \mathbf{x}, \mathbf{v}, u^*) \, du_* \, du^*$$

$$- f(t, \mathbf{x}, \mathbf{v}, u) \int_{D_u} f(t, \mathbf{x}, \mathbf{v}, u^*) \, du^*, \quad (7.6)$$

$$P_i[f] = f(t,\mathbf{x},\mathbf{v},u) \int_{D_u} \eta\,\mu\, f(t,\mathbf{x},\mathbf{v},u^*)\, du^* , \qquad (7.7)$$

where the meaning of the symbols can be recovered by Chapter 6.

The mathematical framework delivered by Eqs. (5.4)-(5.7), already introduced in Chapter 6, corresponds to the assumption that the biological dynamics is ruled by the activity variable only, therefore it corresponds to the spatially homogeneous case perturbed by the velocity jump process modeled by the term L.

The macroscopic equations are derived by asymptotic methods which technically consist in expanding the distribution function f in terms of a small dimensionless parameter ε related to the intermolecular distances (the space-scale parameter) that is equivalent to the connections between the biological constants.

At the basis of the method there is the following hyperbolic scaling

$$\begin{cases} t \to \varepsilon t, \\ \mathbf{x} \to \varepsilon \mathbf{x}, \end{cases} \qquad (7.8)$$

which is equivalent to the choice $t\,v = \frac{1}{\varepsilon}$.

We assume, for the other two biological rates involving this system, that the scaled biological interaction frequency is small compared with the turning frequency and that the (dimensionless) proliferation destruction rate μ is itself small. Accordingly, we will select the following choice of the mechanical and biological constants:

$$v = \frac{1}{\varepsilon}, \qquad \eta = \varepsilon^{q-1}, \qquad \mu = \varepsilon^r, \qquad q \geq 1, r \geq 0. \qquad (7.9)$$

The rigorous scaling produces the following non-dimensional framework:

$$\left(\partial_t + \mathbf{v}\cdot\nabla_{\mathbf{x}}\right) f_\varepsilon = \frac{1}{\varepsilon}\left(L[f_\varepsilon] + \varepsilon^q C[f_\varepsilon] + \varepsilon^{q+r} P[f_\varepsilon]\right). \qquad (7.10)$$

The final target is to understand the asymptotic limit of Eq. (7.10) as ε goes to zero. The limit that is obtained is singular and the convergence properties can be proved under the following technical assumptions.

Assumption 7.1. (Solvability conditions) The turning operator L satisfies, for all f, the following solvability conditions:

$$\int_{D_\mathbf{v}} L[f](t,\mathbf{x},\mathbf{v})\, d\mathbf{v} = 0, \qquad \int_{D_\mathbf{v}} \mathbf{v}L[f](t,\mathbf{x},\mathbf{v})\, d\mathbf{v} = \mathbf{0}. \qquad (7.11)$$

Assumption 7.2. (Kernel of L) For all $\rho \in [0,+\infty)$ and $U \in \mathbb{R}^n$, there exists a unique function

$$M_{\rho,U} = M_{\rho,U}(t,\mathbf{x},\mathbf{v},u) \in L^1(D_\mathbf{v},(1+|\mathbf{v}|)\,d\mathbf{v})$$

such that

$$L[M_{\rho,U}] = 0, \qquad \int_{D_v} M_{\rho,U}\, dv = \rho, \qquad \int_{D_v} \mathbf{v}\, M_{\rho,U}\, dv = \rho\, U, \qquad (7.12)$$

where ρ is the density and U is the mass velocity.

It is worth observing that if we take $\varepsilon = 0$ in (7.10), we formally obtain $L[f_0] = 0$, so f_0 verifies the conditions establish in Assumption 7.2. According to this hypothesis there exists the equilibrium distribution $f_0 = M_{\rho_0, U_0}$, see [Bellomo *et al.* (2007)].

Bearing all the above in mind, we can consider the solution f_ε as a perturbation of the equilibrium M_{ρ_0, U_0} as follows:

$$f_\varepsilon(t, \mathbf{x}, \mathbf{v}, u) = M_{\rho_0, U_0} + \varepsilon g(t, \mathbf{x}, \mathbf{v}, u). \qquad (7.13)$$

The final step is the derivation of the equations satisfied by the equilibrium variables ρ_0 and U_0. According to the paper [Bellomo *et al.* (2007)], we define as a measure of the statistical variation in velocity around the expected mean velocity U_0, the following pressure tensor P_0:

$$P_0(t, \mathbf{x}, u) = \int_{D_v} (\mathbf{v} - U_0) \otimes (\mathbf{v} - U_0)\, f_0\, dv. \qquad (7.14)$$

Thus the following hyperbolic equations with different source terms are obtained:

- $\boxed{r \geq 0, \text{ and } q > 1}$: First order moments with respect to ε generate the hyperbolic system without source term:

$$\begin{cases} \partial_t \rho_0 + \nabla_{\mathbf{x}} \cdot (\rho_0 U_0) = 0, \\[2mm] \partial_t(\rho_0 U_0) + \nabla_{\mathbf{x}} \cdot (\rho_0 U_0 \otimes U_0 + P_0) = 0, \end{cases} \qquad (7.15)$$

which corresponds to *biological tissues characterized by negligible biological (both conservative and proliferative/destructive) activities concerning progression and onset of proliferative phenomena.*

- $\boxed{r > 0, \text{ and } q = 1}$: First order moments with respect to ε yield the following hyperbolic system with a source term related to conservative interactions is obtained:

$$\begin{cases} \partial_t \rho_0 + \nabla_{\mathbf{x}} \cdot (\rho_0 U_0) = \int_{D_v} C(M_{\rho_0, U_0}, M_{\rho_0, U_0})\, dv, \\[2mm] \partial_t(\rho_0 U_0) + \nabla_{\mathbf{x}} \cdot (\rho_0 U_0 \otimes U_0 + P_0) = \int_{D_v} \mathbf{v}\, C(M_{\rho_0, U_0}, M_{\rho_0, U_0})\, dv. \end{cases} \qquad (7.16)$$

This system models *biological tissues at the early stage when cells have initiated mutations, but only conservative biological activities are relevant, while proliferative events have not yet been initiated.*

• $\boxed{r = 0, \text{ and } q = 1}$: In this case the following hyperbolic system with a source term related to both conservative and proliferative interactions is obtained by first order expansion with respect to ε:

$$
\begin{cases}
\partial_t \rho_0 + \nabla_{\mathbf{x}} \cdot (\rho_0 U_0) = \int_{D_{\mathbf{v}}} C(M_{\rho_0, U_0}, M_{\rho_0, U_0}) \, d\mathbf{v} \\[2mm]
\quad + \int_{D_{\mathbf{v}}} P(M_{\rho_0, U_0}, M_{\rho_0, U_0}) \, d\mathbf{v}, \\[2mm]
\partial_t (\rho_0 U_0) + \nabla_{\mathbf{x}} \cdot (\rho_0 U_0 \otimes U_0 + P_0) = \int_{D_{\mathbf{v}}} \mathbf{v} C(M_{\rho_0, U_0}, M_{\rho_0, U_0}) \, d\mathbf{v} \\[2mm]
\quad + \int_{D_{\mathbf{v}}} \mathbf{v} P(M_{\rho_0, U_0}, M_{\rho_0, U_0}) \, d\mathbf{v}.
\end{cases}
\tag{7.17}
$$

This system corresponds to the stage, which generally follows the preceding one, when *both conservative and proliferative/destructive events play a role.*

A different asymptotic method can be developed when the scaling leads to **diffusive models**. For instance, if we choose the following rates

$$
\eta = \varepsilon^q, \quad \mu = \varepsilon^r, \quad q, r \geq 0, \quad \text{and} \quad v = \frac{1}{\varepsilon^p}, \quad p > 0,
\tag{7.18}
$$

and the slow time scale $\tau = \varepsilon t$, the following scaled equation is obtained:

$$
\varepsilon \, \partial_t f_\varepsilon + \mathbf{v} \cdot \nabla_{\mathbf{x}} f_\varepsilon = \frac{1}{\varepsilon^p} L[f_\varepsilon] + \varepsilon^q C[f_\varepsilon] + \varepsilon^{q+r} P[f_\varepsilon].
\tag{7.19}
$$

The analysis is technically different, as documented in [Bellomo and Bellouquid (2004)] and [Bellomo and Bellouquid (2006)]. The macroscopic equations show, in addition to the parabolic structure, some analogy concerning the onset of source terms corresponding to the predominance of one of the three aspects of the biological dynamics, i.e. encounter rate between cells, progressions and proliferative/destructive events, with respect to the other two.

7.3.2 *Asymptotic Methods for Binary Mixtures of Subsystems*

This section deals with the modeling of a binary mixture of multicellular system in biology, thus the kinetic description of the system is encoded by two generalized distribution functions $f_1(t, \mathbf{x}, \mathbf{v}, u)$ and $f_2(t, \mathbf{x}, \mathbf{v}, u)$.

A first generalization of the asymptotic method presented in the last subsection can be technically repeated for the following system

$$
\begin{cases}
(\partial_t + \mathbf{v} \cdot \nabla_{\mathbf{x}}) f_1 = v L[f_1] + C_1[\mathbf{f}, \mathbf{f}] + P_1[\mathbf{f}, \mathbf{f}], \\[2mm]
(\partial_t + \mathbf{v} \cdot \nabla_{\mathbf{x}}) f_2 = v L[f_2] + C_2[\mathbf{f}, \mathbf{f}] + P_2[\mathbf{f}, \mathbf{f}],
\end{cases}
\tag{7.20}
$$

which models the evolution, in the space $\mathbf{x} \in \mathbb{R}^n$ and in the biological state u, of $\mathbf{f} = (f_1, f_2)$. Interactions occur within the action domain of the test cell, which is assumed to be relatively small, so that only binary localized encounters are relevant. Of course, this assumption excludes the possibility of crowding and multiple interactions. Thus the form of the operators L, C_1, C_2, P_1 and P_2 can be generalized as follows:

$$L[f_i] = \int_{D_\mathbf{v}} T(\mathbf{v}^*, \mathbf{v}) f_i(t, \mathbf{x}, \mathbf{v}^*, u) \, d\mathbf{v}^* - f_i(t, \mathbf{x}, \mathbf{v}, u) \int_{D_\mathbf{v}} T(\mathbf{v}, \mathbf{v}^*) d\mathbf{v}^*, \quad (7.21)$$

$$C_i[f_i, f_j] = \sum_{j=1}^{2} \left(\int_{D_u \times D_u} \eta \, \mathscr{B}_{ij}(u_* \to u | u_*, u^*) f_i(t, \mathbf{x}, \mathbf{v}, u_*) f_j(t, \mathbf{x}, \mathbf{v}, u^*) \, du_* \, du^* \right.$$
$$\left. - f_i(t, \mathbf{x}, \mathbf{v}, u) \int_{D_u} f_j(t, \mathbf{x}, \mathbf{v}, u^*) \, du^* \right), \quad (7.22)$$

$$P_i[f_i, f_j] = \sum_{j=1}^{2} f_i(t, \mathbf{x}, \mathbf{v}, u) \int_{D_u} \eta \, \mu \, f_j(t, \mathbf{x}, \mathbf{v}, u^*) \, du^*, \quad (7.23)$$

with obvious meaning of the symbols.

Under the same assumptions stated in the previous subsection, it is possible to prove that $\lim_{\varepsilon \to 0} f_i^\varepsilon = f_i^0 = M_{\rho_i^0, U_i^0}$, for $i \in \{1, 2\}$. Thus, following the same algebra of the previous subsection we are able to derive the following hyperbolic equations with different source terms for the limit density and velocity:

$$\rho_i \equiv \rho_i^0 = \lim_{\varepsilon \to 0} \rho_i^\varepsilon, \quad U_i \equiv U_i^0 = \lim_{\varepsilon \to 0} U_i^\varepsilon,$$

- $\boxed{r \geq 0, \text{ and } q > 1}$:

$$\begin{cases} \partial_t \rho_i + \nabla_\mathbf{x} \cdot (\rho_i U_i) = 0, \\[2mm] \partial_t (\rho_i U_i) + \nabla_\mathbf{x} \cdot (\rho_i U_i \otimes U_i + P_i) = 0, \end{cases} \quad (7.24)$$

- $\boxed{r > 0, \text{ and } q = 1}$:

$$\begin{cases} \partial_t \rho_i + \nabla_\mathbf{x} \cdot (\rho_i U_i) = \int_{D_\mathbf{v}} C_i(M_{\rho_i, U_i}, M_{\rho_i, U_i}) \, d\mathbf{v}, \\[2mm] \partial_t (\rho_i U_i) + \nabla_\mathbf{x} \cdot (\rho_i U_i \otimes U_i + P_i) = \int_{D_\mathbf{v}} \mathbf{v} C_i(M_{\rho_i, U_i}, M_{\rho_i, U_i}) \, d\mathbf{v}, \end{cases} \quad (7.25)$$

- $\boxed{r = 0, \text{ and } q = 1}$:

$$\begin{cases} \partial_t \rho_i + \nabla_{\mathbf{x}} \cdot (\rho_i U_i) = \int_{D_\mathbf{v}} C_i(M_{\rho_i,U_i}, M_{\rho_i,U_i}) \, d\mathbf{v} \\[2mm] \qquad + \int_{D_\mathbf{v}} P_i(M_{\rho_i,U_i}, M_{\rho_i,U_i}) \, d\mathbf{v}, \\[3mm] \partial_t (\rho_i U_i) + \nabla_{\mathbf{x}} \cdot (\rho_i U_i \otimes U_i + P_i) = \int_{D_\mathbf{v}} \mathbf{v} C_i(M_{\rho_i,U_i}, M_{\rho_i,U_i}) \, d\mathbf{v} \\[2mm] \qquad + \int_{D_\mathbf{v}} \mathbf{v} P_i(M_{\rho_i,U_i}, M_{\rho_i,U_i}) \, d\mathbf{v}. \end{cases} \tag{7.26}$$

A further generalization, useful for the applications, is based on the assumption that interactions occur and are weighted, within the action domain $\Omega \subseteq D_{\mathbf{x}}$ of the test particle, by the function $w_{ij}(\mathbf{x}, \mathbf{x}^*)$ that is a normalized (with respect to space integration over Ω) function accounts for the distance and distribution that weakens the intensity of the interaction. The time evolution of the distribution functions vector $\mathbf{f} = (f_1, f_2)$, considering conservative and nonconservative interactions, obeys the following planar system of hyperbolic coupled equations:

$$\begin{cases} (\partial_t + \mathbf{v} \cdot \nabla_{\mathbf{x}}) f_1 = \nu_1 L_1[f_1] + \eta_1 C_1[\mathbf{f}, \mathbf{f}] + \mu_1 P_1[\mathbf{f}, \mathbf{f}], \\[2mm] (\partial_t + \mathbf{v} \cdot \nabla_{\mathbf{x}}) f_2 = \nu_2 L_2[f_2] + \eta_2 C_2[\mathbf{f}, \mathbf{f}] + \mu_2 P_2[\mathbf{f}, \mathbf{f}], \end{cases} \tag{7.27}$$

where:

• The operator $L_i[f_i]$, for $i \in \{1,2\}$, is defined as follows:

$$L_i[f_i] = \int_{D_\mathbf{v}} T_i(\mathbf{v}^*, \mathbf{v}) f_i(t, \mathbf{x}, \mathbf{v}^*, u) \, d\mathbf{v}^* - f_i(t, \mathbf{x}, \mathbf{v}, u) \int_{D_\mathbf{v}} T_i(\mathbf{v}, \mathbf{v}^*) \, d\mathbf{v}^*. \tag{7.28}$$

It is worth mentioning that, more in general, the operator T_i may depend on f_1 and f_2;

• ν_1 and ν_2 denote the mechanical (turning) interaction rates.

• η_1 and η_2 denote the biological interaction rates related to interactions that modify the biological state of the particles of each functional subsystem;

• The operator $C_i = C_i[\mathbf{f}, \mathbf{f}](t, \mathbf{x}, \mathbf{v}, u)$, for $i, j \in \{1, 2\}$, models the gain–loss balance of particles in state u, due to conservative encounters. Accordingly, it has the following form:

$$C_i[\mathbf{f}, \mathbf{f}](t, \mathbf{x}, \mathbf{v}, u) = \sum_{j=1}^{2} C_{ij}[f_i, f_j](t, \mathbf{x}, \mathbf{v}, u), \tag{7.29}$$

where $C_{ij} = C_{ij}[f_i, f_j](t, \mathbf{x}, \mathbf{v}, u)$ reads:

$$C_{ij} = \int_{\Gamma} w_{ij}(\mathbf{x}, \mathbf{x}^*) \mathscr{B}_{ij}(u_* \to u | u_*, u^*) f_i(t, \mathbf{x}, \mathbf{v}, u_*) f_j(t, \mathbf{x}^*, \mathbf{v}, u^*) \, d\mathbf{x}^* \, du_* \, du^*$$

$$- f_i(t, \mathbf{x}, \mathbf{v}, u) \int_{\Lambda} w_{ij}(\mathbf{x}, \mathbf{x}^*) f_j(t, \mathbf{x}^*, \mathbf{v}, u^*) \, d\mathbf{x}^* \, du^*, \tag{7.30}$$

and $\Gamma = \Omega \times D_u \times D_u$, $\Lambda = \Omega \times D_u$, where $\Omega \subseteq D_{\mathbf{x}}$ is the spatial interaction domain.

• The operator $P_i = [\mathbf{f}, \mathbf{f}](t, \mathbf{x}, \mathbf{v}, u)$, for $i, j \in \{1, 2\}$, models net proliferative/destructive interactions into the interacting functional subsystems. This operator is defined as follows:

$$P_i[\mathbf{f}, \mathbf{f}](t, \mathbf{x}, \mathbf{v}, u) = \sum_{j=1}^{2} P_{ij}[f_i, f_j](t, \mathbf{x}, \mathbf{v}, u), \qquad (7.31)$$

where $P_{ij} = P_{ij}[f_i, f_j](t, \mathbf{x}, \mathbf{v}, u)$ reads:

$$P_{ij} = f_i(t, \mathbf{x}, \mathbf{v}, u) \int_{\Lambda} w_{ij}(\mathbf{x}, \mathbf{x}^*) \, \mu_{ij}(u, u^*) \, f_j(t, \mathbf{x}^*, \mathbf{v}, u^*) \, d\mathbf{x}^* du^* . \qquad (7.32)$$

• **The parabolic-parabolic hydrodynamical limit.** The first asymptotic method for this binary mixture deals with parabolic hydrodynamical limits for both functional subsystems with the aim of deriving macroscopic frameworks from the kinetic structure proposed in the system (7.27). These limits depend on the assumptions on the turning operators L_i.

In order to derive macroscopic frameworks we assume the regime $\nu_1 \leq \nu_2$ and also the regime where the biological parameters are small with respect to mechanical ones. After a dimensionless of the system is obtained, a small parameter ε can be chosen such that

$$\nu_1 = \frac{1}{\varepsilon^p}, \quad \nu_2 = \frac{1}{\varepsilon}, \quad p \geq 1,$$

and

$$\eta_1 = \eta_2 = \varepsilon^q, \quad \mu_1 = \varepsilon^{r_1}, \quad \mu_2 = \varepsilon^{r_2},$$

where $q \geq 1$, and r_1, r_2 are non-negative constants.

Some definitions and assumptions are necessary to develop the asymptotic analysis leading to the derivation of macroscopic models. Thus we also assume, in view of the applications of Chapter 9, that the turning operator $L_2[f_2]$ depends on f_1 thanks to the following splitting:

$$L_2[f_1](f_2) = L_2^0[f_2] + \varepsilon L_2^1[f_1](f_2), \qquad (7.33)$$

where L_2^i, for $i \in \{0, 1\}$, is given by

$$L_2^i[f_2] = \int_{D_{\mathbf{v}}} \left[T_2^i(\mathbf{v}, \mathbf{v}^*) f_2(t, \mathbf{x}, \mathbf{v}^*, u) - T_2^i(\mathbf{v}^*, \mathbf{v}) f_2(t, \mathbf{x}, \mathbf{v}, u) \right] d\mathbf{v}^* . \qquad (7.34)$$

We also suppose that L_2^0 is independent of f_1.

Accordingly, the model (7.27) can be written in the following form:

$$\begin{cases} (\varepsilon \partial_t + \mathbf{v} \cdot \nabla_{\mathbf{x}}) f_1^\varepsilon = \frac{1}{\varepsilon^p} L[f_1^\varepsilon] + \varepsilon^q C_1[\mathbf{f}^\varepsilon, \mathbf{f}^\varepsilon] + \varepsilon^{q+r_1} P_1[\mathbf{f}^\varepsilon, \mathbf{f}^\varepsilon], \\[2ex] (\varepsilon \partial_t + \mathbf{v} \cdot \nabla_{\mathbf{x}}) f_2^\varepsilon = \frac{1}{\varepsilon} L_2[f_1^\varepsilon](f_2^\varepsilon) + \varepsilon^q C_2[\mathbf{f}^\varepsilon, \mathbf{f}^\varepsilon] + \varepsilon^{q+r_2} P_2[\mathbf{f}^\varepsilon, \mathbf{f}^\varepsilon], \end{cases} \qquad (7.35)$$

where $\mathbf{f}^\varepsilon = (f_1^\varepsilon, f_2^\varepsilon)$.

In what follows, the integral with respect to the variable $\mathbf{v}$ will be denoted by $\langle \cdot \rangle_{\mathbf{v}}$.

Assumption 7.3. We assume that the turning operators L_1 and L_2 satisfy, for all g, the following conditions:

$$\int_{D_{\mathbf{v}}} L_1[g]d\mathbf{v} = \int_{D_{\mathbf{v}}} L_2^0[g]d\mathbf{v} = \int_{D_{\mathbf{v}}} L_2^1[f_1](g)d\mathbf{v} = 0. \tag{7.36}$$

Assumption 7.4. There exists a bounded velocity distribution $M_i(\mathbf{v}) > 0$, for $i \in \{1, 2\}$, independent of $t, \mathbf{x}$, such that the detailed balance

$$T_1(\mathbf{v}, \mathbf{v}^*)M_1(\mathbf{v}^*) = T_1(\mathbf{v}^*, \mathbf{v})M_1(\mathbf{v}) \tag{7.37}$$

and

$$T_2^0(\mathbf{v}, \mathbf{v}^*)M_2(\mathbf{v}^*) = T_2^0(\mathbf{v}^*, \mathbf{v})M_2(\mathbf{v}) \tag{7.38}$$

holds. Moreover, the flow produced by these equilibrium distributions vanishes, and M_i are normalized, i.e. $\langle \mathbf{v} M_i(\mathbf{v}) \rangle_{\mathbf{v}} = 0$ and $\langle M_i(\mathbf{v}) \rangle_{\mathbf{v}} = 1$.

Assumption 7.5. The kernels $T_1(\mathbf{v}, \mathbf{v}^*)$ and $T_2^0(\mathbf{v}, \mathbf{v}^*)$ are bounded, and there exists a constant $\sigma_i > 0$, $i = 1, 2$, such that

$$T_1(\mathbf{v}, \mathbf{v}^*) \geq \sigma_1 M_1(\mathbf{v}), \quad T_2^0(\mathbf{v}, \mathbf{v}^*) \geq \sigma_2 M_2(\mathbf{v}), \tag{7.39}$$

for all $(\mathbf{v}, \mathbf{v}^*) \in D_{\mathbf{v}} \times D_{\mathbf{v}}$, $\mathbf{x} \in \Omega$, and $t > 0$.

Assumption 7.6. There exist C_i, $i = 1, 2, 3$ independent of t, $\mathbf{x}$, and $\mathbf{v}$ such that:

$$T_1(\mathbf{v}, \mathbf{v}^*) \leq C_1 M_1(\mathbf{v}), \quad T_2^0(\mathbf{v}, \mathbf{v}^*) \leq C_2 M_2(\mathbf{v}), \quad |T_2^1[f_1]| \leq C_3 |f_1|.$$

Letting $T_1 = L_1$ and $T_2 = L_2^0$, the above assumptions yields the following properties (see [Bellomo and Bellouquid (2006)]):

i) For $f \in L^2$, the equation $T_i[g] = f$, for $i \in \{1, 2\}$, has a unique solution

$$g \in L^2\left(D_{\mathbf{v}}, \frac{d\mathbf{v}}{M_i}\right),$$

which satisfies

$$\langle g \rangle_{\mathbf{v}} = \int_{D_{\mathbf{v}}} g(\mathbf{v}) \, d\mathbf{v} = 0 \quad \text{if and only if} \quad \langle f \rangle_{\mathbf{v}} = \int_{D_{\mathbf{v}}} f(\mathbf{v}) \, d\mathbf{v} = 0.$$

ii) The operator T_i is self-adjoint in the space $L^2\left(D_{\mathbf{v}}, \frac{d\mathbf{v}}{M_i}\right)$.

iii) The equation $T_i[g] = \mathbf{v} M_i(\mathbf{v})$, $i = 1, 2$, has a unique solution that we call $\theta_i(\mathbf{v})$.

iv) The kernel of T_i is $N(L_i) = vect(M_i(\mathbf{v}))$, $i \in \{1, 2\}$.

The limit $\varepsilon \to 0$ is formally developed, in this subsection, for (7.35). The resulting macroscopic framework depends on the properties of the turning operators. The strategy to derive the macroscopic model consists in the following steps:

Step 1. Multiplying the first equation of (7.35) by ε^p and letting ε go to zero, yields $L_1[f_1^0] = 0$. Therefore, one deduces, by iv) that there exists a function S, independent of $\mathbf{v}$, such that

$$f_1^0(t,\mathbf{x},\mathbf{v},u) = S(t,\mathbf{x},u)M_1(\mathbf{v}). \tag{7.40}$$

By multiplying the second equation of (7.35) by ε, using (7.33), and letting ε go to zero, yields $L_2^0[f_2^0] = 0$. Moreover, analogous reasonings yield

$$f_2^0(t,\mathbf{x},\mathbf{v},u) = n(t,\mathbf{x},u)M_2(\mathbf{v}). \tag{7.41}$$

Step 2. Integration of the first and the second equations in (7.35) over $\mathbf{v}$ and using (7.36) yields:

$$\partial_t \langle f_1^\varepsilon \rangle_\mathbf{v} + \frac{1}{\varepsilon} \langle \mathbf{v} \cdot \nabla_\mathbf{x} f_1^\varepsilon \rangle_\mathbf{v} = \varepsilon^{q-1} \langle C_1[\mathbf{f}^\varepsilon, \mathbf{f}^\varepsilon] \rangle_\mathbf{v} + \varepsilon^{q+r_1-1} \langle P_1[\mathbf{f}^\varepsilon, \mathbf{f}^\varepsilon] \rangle_\mathbf{v}, \tag{7.42}$$

and

$$\partial_t \langle f_2^\varepsilon \rangle_\mathbf{v} + \frac{1}{\varepsilon} \langle \mathbf{v} \cdot \nabla_\mathbf{x} f_2^\varepsilon \rangle_\mathbf{v} = \varepsilon^{q-1} \langle C_2[\mathbf{f}^\varepsilon, \mathbf{f}^\varepsilon] \rangle_\mathbf{v} + \varepsilon^{q+r_2-1} \langle P_2[\mathbf{f}^\varepsilon, \mathbf{f}^\varepsilon] \rangle_\mathbf{v}. \tag{7.43}$$

The asymptotic limit of $\frac{1}{\varepsilon} \langle \mathbf{v} \cdot \nabla_\mathbf{x} f_i^\varepsilon \rangle_\mathbf{v}$, $i = 1,2$, needs to be estimated to recover the limit in (7.42) and (7.43). After simple algebra and thanks to the properties that L_2^0 and L_1 are self-adjoint operators, one deduces that

$$\lim_{\varepsilon \to 0} \frac{1}{\varepsilon} \langle \mathbf{v} \cdot \nabla_\mathbf{x} f_1^\varepsilon \rangle_\mathbf{v} = \begin{cases} \mathrm{div}_\mathbf{x} \langle \theta_1 (\mathbf{v} \cdot \nabla_\mathbf{x}) S \rangle_\mathbf{v} & \text{if } p = 1 \\ 0 & \text{if } p > 1 \end{cases}$$

and

$$\lim_{\varepsilon \to 0} \frac{1}{\varepsilon} \langle \mathbf{v} \cdot \nabla_\mathbf{x} f_2^\varepsilon \rangle_\mathbf{v} = \mathrm{div}_\mathbf{x} \langle \theta_2 (\mathbf{v} \cdot \nabla_\mathbf{x} n) \rangle_\mathbf{v} - \mathrm{div}_\mathbf{x} \left\langle \frac{\theta_2}{M_2(\mathbf{v})} L_2^1 [M_1 S](M_2 n) \right\rangle_\mathbf{v}.$$

The asymptotic quadratic terms of (7.42) and (7.43) converge, for $i \in \{1,2\}$, to the following functionals:

$$\mathscr{C}_i(n,S)(t,\mathbf{x},u) = \left\langle C_i \left[\begin{pmatrix} M_1 S \\ M_2 n \end{pmatrix}, \begin{pmatrix} M_1 S \\ M_2 n \end{pmatrix} \right] \right\rangle,$$

and

$$\mathscr{P}_i(n,S)(t,\mathbf{x},u) = \left\langle P_i \left[\begin{pmatrix} M_1 S \\ M_2 n \end{pmatrix}, \begin{pmatrix} M_1 S \\ M_2 n \end{pmatrix} \right] \right\rangle.$$

Therefore, we can derive macroscopic models by taking limits in (7.42) and (7.43). To pass to the limit it is sufficient to assume pointwise convergence together with a global L^p bound of f_i^ε. This result can be stated as follows.

Theorem 7.1. *Let $f_i^\varepsilon(t, \mathbf{x}, \mathbf{v}, u)$ be a sequence of solutions to the scaled kinetic system (7.35), which verifies Assumptions 7.3-7.6 such that f_i^ε converges a.e. in $[0, \infty) \times \Omega \times D_\mathbf{v} \times D_u$ to a function f_i^0 as ε goes to zero and*

$$\sup_{t \geq 0} \int_\Omega \int_{D_\mathbf{v}} \int_{D_u} |f_i^\varepsilon(t, \mathbf{x}, \mathbf{v}, u)|^m \, du \, d\mathbf{v} \, d\mathbf{x} \leq C < \infty \qquad (7.44)$$

for some positive constants $C > 0$ and $m > 2$. Moreover, we assume that the probability kernels $\mathscr{B}_{ij}$ are bounded functions and that the weight functions w_{ij} have finite integrals. It follows that the asymptotic limits f_i^0 have the form (7.40)-(7.41) where n, S are the weak solutions of the following equation (that depends on the values of p, q, r_1 and r_2)

$$\partial_t S - \delta_{p,1} \operatorname{div}_\mathbf{x}(D_S \cdot \nabla_\mathbf{x} S) = \delta_{q,1} \mathscr{C}_1(n, S) + \delta_{q,1} \delta_{r_1,0} \mathscr{P}_1(n, S),$$

$$\partial_t n + \operatorname{div}_\mathbf{x}(n \, \alpha(S) - D_n \cdot \nabla_\mathbf{x} n) = \delta_{q,1} \mathscr{C}_2(n, S) + \delta_{q,1} \delta_{r_2,0} \mathscr{P}_2(n, S),$$

where $\delta_{a,b}$ stands for the Kronecker delta and D_n, D_S and $\alpha(S)$ are given by

$$D_S = -\int_{D_\mathbf{v}} \mathbf{v} \otimes \theta_1(\mathbf{v}) d\mathbf{v}, \qquad D_n = -\int_{D_\mathbf{v}} \mathbf{v} \otimes \theta_2(\mathbf{v}) d\mathbf{v} \qquad (7.45)$$

and

$$\alpha(S) = -\int_{D_\mathbf{v}} \frac{\theta_2(\mathbf{v})}{M_2(\mathbf{v})} L_2^1[M_1 S](M_2)(\mathbf{v}) d\mathbf{v}. \qquad (7.46)$$

The approach we have developed is quite general. Some more specific examples are given in Chapter 9.

• **The parabolic-hyperbolic scaling.** The asymptotic analysis previously developed was based on the assumption of the *parabolic scaling* for both the functional subsystems. The limit gives rise to a system of coupled equations which include a diffusion term for both subsystems. Assuming that the second subsystem has no diffusive behavior, the derivation of macroscopic equations requires a *hyperbolic scaling* for this population. Bearing this in mind, let us now consider the system (7.27) with a parabolic scaling for the first subsystem, but with an hyperbolic one for the second one:

$$(\varepsilon \partial_t + \mathbf{v} \cdot \nabla_\mathbf{x}) f_1^\varepsilon = \frac{1}{\varepsilon^p} L_1[f_1^\varepsilon] + \varepsilon^q C_1[\mathbf{f}^\varepsilon, \mathbf{f}^\varepsilon] + \varepsilon^{q+r_1} P_1[\mathbf{f}^\varepsilon, \mathbf{f}^\varepsilon], \qquad (7.47)$$

$$\varepsilon(\partial_t + \mathbf{v} \cdot \nabla_\mathbf{x}) f_2^\varepsilon = L_2[f_1^\varepsilon](f_2^\varepsilon) + \varepsilon^q C_2[\mathbf{f}^\varepsilon, \mathbf{f}^\varepsilon] + \varepsilon^{q+r_2} P_2[\mathbf{f}^\varepsilon, \mathbf{f}^\varepsilon], \qquad (7.48)$$

where $p, q \geq 1$, $r_1, r_2 \geq 0$, and ε is a small parameter that is allowed to tend to zero.

Considering that the scaling for the first equation (7.47) is exactly that of previous asymptotic method, the hypothesis on L_1 and the passage to the limit follow the same guidelines. This approach yields:

$$\partial_t S = \delta_{p,1} \operatorname{div}_{\mathbf{x}} (D_S \cdot \nabla_{\mathbf{x}} S) + \delta_{q,1} \left\langle C_1 \left[\begin{pmatrix} M_1 S \\ f_2^0 \end{pmatrix}, \begin{pmatrix} M_1 S \\ f_2^0 \end{pmatrix} \right] \right\rangle$$

$$+ \delta_{q,1}, \delta_{r_1,0} \left\langle P_1 \left[\begin{pmatrix} M_1 S \\ f_2^0 \end{pmatrix}, \begin{pmatrix} M_1 S \\ f_2^0 \end{pmatrix} \right] \right\rangle,$$

where $\delta_{a,b}$ stands for the Kronecker delta and f_2^0 will be given by the limit of f_2^ε, to be determined.

Actually, some different "hyperbolic" hypothesis on the operator $L_2[f_1^\varepsilon]$ are required to establish the behavior of the second subsystem and its macroscopical limit.

We assume that the turning operator $L_2[f_1^\varepsilon]$ is decomposed as in (7.33) and verify (7.34), meanwhile condition (7.36) and Assumptions 7.4 and 7.5 are replaced by the following hyperbolic Assumptions.

Assumption 7.7. The turning operator $L_2[f_1^\varepsilon] = L_2^0 + \varepsilon L_2^1[f_1^\varepsilon]$ satisfies the following conditions:

$$\int_{D_{\mathbf{v}}} L_2^0[g] d\mathbf{v} = \int_{D_{\mathbf{v}}} L_2^1[f_1](g) d\mathbf{v} = 0, \tag{7.49}$$

$$\int_{D_{\mathbf{v}}} \mathbf{v} L_2^0[g] d\mathbf{v} = 0. \tag{7.50}$$

Assumption 7.8. For any $n \in [0, +\infty)$ and $U \in \mathbb{R}^n$, there exists a unique function $M_{n,U} \in L^1(D_{\mathbf{v}}, (1 + |\mathbf{v}|) d\mathbf{v})$ such that

$$L_2^0[M_{n,U}] = 0, \quad \int_{D_{\mathbf{v}}} M_{n,U}(v) d\mathbf{v} = n, \quad \int_{D_{\mathbf{v}}} \mathbf{v} M_{n,U}(\mathbf{v}) d\mathbf{v} = nU. \tag{7.51}$$

Then, we let ε going to zero in Eq. (7.48). This yields $L_2^0[f_2^0] = 0$. Therefore, as a consequence, there exist $n \geq 0$ and $U \in \mathbb{R}^n$ (depending on $(t, \mathbf{x}, u)$), namely the macroscopic density and velocity associated to function f_2^0, such that $f_2^0 = M_{n,U}$.

The next step consists in determining the macroscopic dynamics for n and U and the coupling with the macroscopic density S. To do that, we integrate (7.48) over $\mathbf{v}$ and use (7.49) to obtain

$$\partial_t \langle f_2^\varepsilon \rangle_{\mathbf{v}} + \langle \mathbf{v} \cdot \nabla_{\mathbf{x}} f_2^\varepsilon \rangle_{\mathbf{v}} = \varepsilon^{q-1} \langle C_2[\mathbf{f}^\varepsilon, \mathbf{f}^\varepsilon] \rangle_{\mathbf{v}} + \varepsilon^{q-1+r_2} \langle P_2[\mathbf{f}^\varepsilon, \mathbf{f}^\varepsilon] \rangle_{\mathbf{v}}.$$

Now, we find, by letting $\varepsilon \to 0$, that the function n satisfies the following conservation law

$$\frac{\partial n}{\partial t} + \operatorname{div}_{\mathbf{x}}(nU) = \delta_{q,1} \langle C_2[\mathbf{f}^0, \mathbf{f}^0] \rangle_{\mathbf{v}} + \delta_{q,1} \delta_{r_2,0} \langle P_2[\mathbf{f}^0, \mathbf{f}^0] \rangle_{\mathbf{v}}$$

at the equilibrium, where

$$\mathbf{f}^0 = (M_1(\mathbf{v})\,S, M_{n,U}). \tag{7.52}$$

In the same way, multiplying (7.48) by $\mathbf{v}$, integrating over $\mathbf{v}$, and using (7.50) yields

$$\partial_t \langle \mathbf{v} f_2^\varepsilon \rangle_\mathbf{v} + \mathrm{Div}_\mathbf{x} \langle \mathbf{v} \otimes \mathbf{v} f_2^\varepsilon \rangle_\mathbf{v} = \langle \mathbf{v} L_2^1 [f_1^\varepsilon](f_2^\varepsilon) \rangle_\mathbf{v}$$
$$+ \, \varepsilon^{q-1} \langle \mathbf{v} C_2[\mathbf{f}^\varepsilon, \mathbf{f}^\varepsilon] \rangle_\mathbf{v} + \varepsilon^{q-1+r_2} \langle \mathbf{v} P_2[\mathbf{f}^\varepsilon, \mathbf{f}^\varepsilon] \rangle_\mathbf{v}.$$

By letting again $\varepsilon \to 0$, the limit equation for the momentum is rapidly obtained:

$$\partial_t(nU) + \mathrm{Div}\,(nU \otimes U + \mathbb{P}) = \langle \mathbf{v} L_2^1 [M_2\,S](M_{n,U}) \rangle_\mathbf{v}$$
$$+ \, \delta_{q,1} \langle \mathbf{v} C_2[\mathbf{f}^0, \mathbf{f}^0] \rangle_\mathbf{v} + \delta_{q,1} \delta_{r_2,0} \langle \mathbf{v} P_2[\mathbf{f}^0, \mathbf{f}^0] \rangle_\mathbf{v},$$

where $\mathbf{f}^0$ is defined by (7.52) and the pressure tensor is, as usual, defined by

$$\mathbb{P} = \int_{D_\mathbf{v}} (\mathbf{v} - U) \otimes (\mathbf{v} - U) M_{n,U}\, d\mathbf{v}. \tag{7.53}$$

We can summarize this result in the following theorem.

Theorem 7.2. *Let $f_i^\varepsilon(t, \mathbf{x}, \mathbf{v}, u)$ be a sequence of solutions to the scaled kinetic system (7.47)-(7.48) with L_1 verifying Assumptions 7.3-7.6 and L_2 verifying Assumptions 7.7 and 7.8. Assume that f_i^ε verifies (7.44) and converges a.e. in $[0, \infty) \times \Omega \times D_\mathbf{v} \times D_u$ to a function f_i^0 as ε goes to zero. Moreover, we assume that the probability kernels $\mathscr{B}_{ij}$ are bounded functions and that the weight functions w_{ij} have finite integrals. Then, the asymptotic limit is given by (7.52) where S, n and U are weak solutions of the following system:*

$$\begin{cases} \partial_t S = \delta_{p,1} \mathrm{div}_\mathbf{x}\,(D_S \cdot \nabla_\mathbf{x} S) + \delta_{q,1} \langle C_1[\mathbf{f}^0, \mathbf{f}^0] \rangle_\mathbf{v} + \delta_{q,1} \delta_{r_1,0} \langle P_1[\mathbf{f}^0, \mathbf{f}^0] \rangle_\mathbf{v}, \\[2mm] \partial_t n + \mathrm{div}_\mathbf{x}\,(nU) = \delta_{q,1} \langle C_2[\mathbf{f}^0, \mathbf{f}^0] \rangle_\mathbf{v} + \delta_{q,1} \delta_{r_2,0} \langle P_2[\mathbf{f}^0, \mathbf{f}^0] \rangle_\mathbf{v}, \\[2mm] \partial_t(nU) + \mathrm{Div}(nU \otimes U + \mathbb{P}) = \langle \mathbf{v} L_2^1 [M_1\,S](M_{n,U}) \rangle_\mathbf{v} \\[2mm] \qquad\qquad + \delta_{q,1} \langle \mathbf{v} C_2[\mathbf{f}^0, \mathbf{f}^0] \rangle_\mathbf{v} + \delta_{q,1} \delta_{r_2,0} \langle \mathbf{v} P_2[\mathbf{f}^0, \mathbf{f}^0] \rangle_\mathbf{v}. \end{cases} \tag{7.54}$$

Remark 7.1. Note that the influence of the population S on the velocity U is given as an integral source term. Moreover, even if we take $L_2^1 = 0$, the other integral terms give an analogous coupling once $q = 1$.

7.4   Cellular-Tissue Scale Modeling of Open Systems

Let us now consider the approach for an open system, namely subject to external (possibly therapeutical) action for several interacting functional subsystems, where mutations induce the generation of new subsystems. This topic, which is an interesting interplay with real biological applications has been dealt with in [Bellomo *et al.* (2010)], which is the main reference for the survey of this section. The scaling selected for the asymptotic analysis is again that hyperbolic.

The analysis is developed referring to the mathematical structure defined in Eqs. (5.14)-(5.18). The hyperbolic scaling yields:

$$\varepsilon \left(\partial_t + \mathbf{v} \cdot \nabla_{\mathbf{x}} \right) f_i^\varepsilon = L_i[f_i^\varepsilon] + \varepsilon^{q_i} J_i^\varepsilon [\mathbf{f}^\varepsilon] + \varepsilon \, Q_i^\varepsilon [\mathbf{f}^\varepsilon, \mathbf{g}], \qquad (7.55)$$

for the perturbation f_i^ε and $i \in \{1, 2, \dots, n\}$.

Here, the scaled turning operator $L_i[f_i]$ is considered to have the same expression (7.28), but with non-dimensional constants. The interaction operator for the closed system is scaled as follows:

$$J_i^\varepsilon [\mathbf{f}^\varepsilon] = \sum_{j=1}^{n} \left(G_{ij}[\mathbf{f}^\varepsilon] - L_{ij}[\mathbf{f}^\varepsilon] \right) (t, \mathbf{x}, \mathbf{v}, u) + \varepsilon^{r_i} \sum_{h=1}^{n} \sum_{k=1}^{n} P_{hk}^i [\mathbf{f}^\varepsilon],$$

and

$$Q_i^\varepsilon [\mathbf{f}^\varepsilon] = \sum_{j=1}^{n} \left(G_{ij}^e[\mathbf{f}^\varepsilon] - L_{ij}^e[\mathbf{f}^\varepsilon] \right) (t, \mathbf{x}, \mathbf{v}, u) + \varepsilon^{r_i} \sum_{j=1}^{n} \sum_{j=1}^{m} P_{ij}^e [\mathbf{f}^\varepsilon],$$

where we have retained the same notation for the non-dimensional gain G_{ij}, lost L_{ij} and proliferative/destructive P_{jk}^i term. Finally, the external therapeutical action preserves its own form although it is multiplied by ε.

The hyperbolic macroscopic behavior is deduced from the limit $\varepsilon \to 0$. First, taking $\varepsilon = 0$ one formally obtains $L_i[f_i^0] = 0$, so each f_i^0 verifies the conditions of Assumption 7.8. Therefore, the limiting n distributions of the form $f_i^0 = M_{\rho_i^0, \mathbf{U}_i^0}$ are obtained corresponding to each subsystem. The equations satisfied by the equilibrium variables ρ_i^0 and $\mathbf{U}_i^0$ have to be studied. Integration over $\mathbf{v}$ in (7.55) yields:

$$\partial_t \rho_i^\varepsilon + \operatorname{div}(\rho_i^\varepsilon \mathbf{U}_i^\varepsilon)$$

$$= \varepsilon^{q_i-1} \sum_{j=1}^{n} \int_{D_{\mathbf{v}}} \left(G_{ij}[\mathbf{f}^\varepsilon] - L_{ij}[\mathbf{f}^\varepsilon] \right) d\mathbf{v} + \varepsilon^{q_i+r_i-1} \sum_{h=1}^{n} \sum_{k=1}^{n} \int_{D_{\mathbf{v}}} P_{hk}^i [\mathbf{f}^\varepsilon] d\mathbf{v}$$

$$+ \varepsilon^{q_i-1} \sum_{j=1}^{m} \int_{D_{\mathbf{v}}} \left(G_{ij}^e[\mathbf{f}^\varepsilon] - L_{ij}^e[\mathbf{f}^\varepsilon] \right) d\mathbf{v} + \varepsilon^{q_i+r_i-1} \sum_{j=1}^{m} \int_{D_{\mathbf{v}}} P_{ij}^e [\mathbf{f}^\varepsilon, \mathbf{g}] d\mathbf{v}.$$

Analogously, multiplying (7.55) by $\mathbf{v}$ and integrating leads to

$$\partial_t(\rho_i^\varepsilon \mathbf{U}_i^\varepsilon) + \text{Div}\left(\int_{D_\mathbf{v}} \mathbf{v} \otimes \mathbf{v} f_i^\varepsilon \, d\mathbf{v}\right)$$

$$= \varepsilon^{q_i-1} \sum_{j=1}^n \int_{D_\mathbf{v}} \mathbf{v}\left(G_{ij}[\mathbf{f}^\varepsilon] - L_{ij}[\mathbf{f}^\varepsilon]\right) d\mathbf{v} + \varepsilon^{q_i+r_i-1} \sum_{h=1}^n \sum_{k=1}^n \int_{D_\mathbf{v}} \mathbf{v} P_{hk}^i[\mathbf{f}^\varepsilon] d\mathbf{v}$$

$$+ \varepsilon^{q_i-1} \sum_{j=1}^m \int_{D_\mathbf{v}} \mathbf{v}\left(G_{ij}^e[\mathbf{f}^\varepsilon] - L_{ij}^e[\mathbf{f}^\varepsilon]\right) d\mathbf{v} + \varepsilon^{q_i+r_i-1} \sum_{j=1}^m \int_{D_\mathbf{v}} \mathbf{v} P_{ij}^e[\mathbf{f}^\varepsilon, \mathbf{g}] d\mathbf{v}.$$

Let us now consider a solution as a perturbation of the equilibrium $M_{\rho_i^0, \mathbf{U}_i^0}^i$ in the following form:

$$f_i = M_{\rho_i^0, \mathbf{U}_i^0}^i + \varepsilon h_i.$$

Denoting $\mathbf{M} = \{M_{\rho_i^0, \mathbf{U}_i^0}^i\}_{i=1}^n$ yields

$$\partial_t \rho_i^0 + \text{div}(\rho_i^0 \mathbf{U}_i^0) = O(\varepsilon^{q_i})$$

$$+ \varepsilon^{q_i-1} \sum_{j=1}^n \int_{D_\mathbf{v}} \left(G_{ij}[\mathbf{M}] - L_{ij}[\mathbf{M}]\right) d\mathbf{v} + \varepsilon^{q_i+r_i-1} \sum_{h=1}^n \sum_{k=1}^n \int_{D_\mathbf{v}} P_{hk}^i[\mathbf{M}] d\mathbf{v}$$

$$+ \varepsilon^{q_i-1} \sum_{j=1}^m \int_{D_\mathbf{v}} \left(G_{ij}^e[\mathbf{M}, \mathbf{g}] - L_{ij}^e[\mathbf{M}, \mathbf{g}]\right) d\mathbf{v} + \varepsilon^{q_i+r_i-1} \sum_{j=1}^m \int_{D_\mathbf{v}} P_{ij}^e[\mathbf{M}, \mathbf{g}] d\mathbf{v},$$

and

$$\partial_t(\rho_i^0 \mathbf{U}_i^0) + \text{Div}\left(\int_{D_\mathbf{v}} \mathbf{v} \otimes \mathbf{v} M_{\rho_i^0, \mathbf{U}_i^0} d\mathbf{v}\right) = O(\varepsilon^{q_i})$$

$$+ \varepsilon^{q_i-1} \sum_{j=1}^n \int_{D_\mathbf{v}} \mathbf{v}\left(G_{ij}[\mathbf{M}] - L_{ij}[\mathbf{M}]\right) d\mathbf{v} + \varepsilon^{q_i+r_i-1} \sum_{h=1}^n \sum_{k=1}^n \int_{D_\mathbf{v}} \mathbf{v} P_{hk}^i[\mathbf{M}] d\mathbf{v}$$

$$+ \varepsilon^{q_i-1} \sum_{j=1}^m \int_{D_\mathbf{v}} \mathbf{v}\left(G_{ij}^e[\mathbf{M}, \mathbf{g}] - L_{ij}^e[\mathbf{M}, \mathbf{g}]\right) d\mathbf{v} + \varepsilon^{q_i+r_i-1} \sum_{j=1}^m \int_{D_\mathbf{v}} \mathbf{v} P_{ij}^e[\mathbf{M}, \mathbf{g}] d\mathbf{v}.$$

Let us now define, as usual, the pressure tensor P_i^0 as a measure of the statistical variation in velocity around the expected mean velocity $\mathbf{U}_i^0$,

$$P_i^0(t, x, u) = \int_{D_\mathbf{v}} (\mathbf{v} - \mathbf{U}_i^0) \otimes (\mathbf{v} - \mathbf{U}_i^0) f_i^0 \, d\mathbf{v}. \tag{7.56}$$

Rewriting (7.56) as follows

$$\int_{D_\mathbf{v}} \mathbf{v} \otimes \mathbf{v} M_{\rho_i^0, \mathbf{U}_i^0} d\mathbf{v} = P_i^0 + \rho_i^0 (\mathbf{U}_i^0 \otimes \mathbf{U}_i^0), \tag{7.57}$$

is useful to eliminate $\mathbf{v}$ in the second equation.

Some specific cases, which measure the relation between the mechanical variables, the biological rates and the therapies can be considered depending on the scales of the biological functions. Different hyperbolic systems are obtained:

Case 1. $\boxed{r_i \geq 0 \text{ and } q_i > 1}$. *This system has no source terms considering that both progression and proliferative phenomena are not yet relevant:* The model consists in the following conservative hyperbolic system:

$$
\begin{cases}
\partial_t \rho_i^0 + \mathrm{div}(\rho_i^0 \mathbf{U}_i^0) = 0, \\[2mm]
\partial_t (\rho_i^0 \mathbf{U}_i^0) + \mathrm{Div}\big(\rho_i^0 (\mathbf{U}_i^0 \otimes \mathbf{U}_i^0) + P^0\big) = 0.
\end{cases}
\tag{7.58}
$$

Case 2. $\boxed{r_i > 0 \text{ and } q_i = 1}$. *In this case a source term is preserved related to conservative actions, and therapy actions into the closed system:*

$$
\begin{cases}
\partial_t \rho_i^0 + \mathrm{div}(\rho_i^0 \mathbf{U}_i^0) = \displaystyle\sum_{j=1}^{n} \int_{D_{\mathbf{v}}} \Big(G_{ij}[\mathbf{M}] - L_{ij}[\mathbf{M}]\Big)\, d\mathbf{v} \\[2mm]
\qquad + \displaystyle\sum_{j=1}^{m} \int_{D_{\mathbf{v}}} \Big(G_{ij}^e[\mathbf{M},\mathbf{g}] - L_{ij}^e[\mathbf{M},\mathbf{g}]\Big)\, d\mathbf{v}, \\[4mm]
\partial_t (\rho_i^0 \mathbf{U}_i^0) + \mathrm{Div}\big(\rho_i^0 (\mathbf{U}_i^0 \otimes \mathbf{U}_i^0) + P^0\big) = \displaystyle\sum_{j=1}^{n} \int_{D_{\mathbf{v}}} \mathbf{v}\Big(G_{ij}[\mathbf{M}] - L_{ij}[\mathbf{M}]\Big)\, d\mathbf{v} \\[2mm]
\qquad + \displaystyle\sum_{j=1}^{m} \int_{D_{\mathbf{v}}} \mathbf{v}\Big(G_{ij}^e[\mathbf{M},\mathbf{g}] - L_{ij}^e[\mathbf{M},\mathbf{g}]\Big)\, d\mathbf{v}.
\end{cases}
\tag{7.59}
$$

Case 3. $\boxed{r_i = 0 \text{ and } q_i = 1}$. *In this last case all the macroscopic information about the closed system is preserved, including proliferative, destructive interactions, and therapy actions:*

$$
\begin{cases}
\partial_t \rho_i^0 + \mathrm{div}(\rho_i^0 \mathbf{U}_i^0) = \displaystyle\sum_{j=1}^{n} \int_{D_{\mathbf{v}}} \Big(G_{ij}[\mathbf{M}] - L_{ij}[\mathbf{M}]\Big)\, d\mathbf{v} + \sum_{h=1}^{n}\sum_{k=1}^{n} P_{hk}^i[\mathbf{M}]\, d\mathbf{v} \\[2mm]
\qquad + \displaystyle\sum_{j=1}^{m} \int_{D_{\mathbf{v}}} \Big(G_{ij}^e[\mathbf{M},\mathbf{g}] - L_{ij}^e[\mathbf{M},\mathbf{g}] + P_{ij}^e[\mathbf{M},\mathbf{g}]\Big)\, d\mathbf{v}, \\[4mm]
\partial_t (\rho_i^0 \mathbf{U}_i^0) + \mathrm{Div}\big(\rho_i^0 (\mathbf{U}_i^0 \otimes \mathbf{U}_i^0) + P^0\big) \\[2mm]
\qquad = \displaystyle\sum_{j=1}^{n} \int_{D_{\mathbf{v}}} \mathbf{v}\Big(G_{ij}[\mathbf{M},\mathbf{g}] - L_{ij}[\mathbf{M},\mathbf{g}]\Big)\, d\mathbf{v} + \sum_{h=1}^{n}\sum_{k=1}^{n} S_{hk}^i[\mathbf{M}]\Big)\, d\mathbf{v} \\[2mm]
\qquad + \displaystyle\sum_{j=1}^{m} \int_{D_{\mathbf{v}}} \mathbf{v}\Big(G_{ij}^e[\mathbf{M},\mathbf{g}] - L_{ij}^e[\mathbf{M},\mathbf{g}] + P_{ij}^e[\mathbf{M},\mathbf{g}]\Big)\, d\mathbf{v}.
\end{cases}
\tag{7.60}
$$

Remark 7.1. The approach we have presented here is quite general. It is easy to include some simple examples by choosing a concrete turning operator for which $\mathbf{M}$ has an explicit form (see Chapter 9).

The convergence properties of the mathematical approach have been studied in [Bellomo *et al.* (2010)] and can be stated in the following theorem:

Theorem *Let* $\mathbf{f}^\varepsilon$ *be a solution of (7.55) verifying*

$$\|\mathbf{f}^\varepsilon\|_{C(0,\infty;L^p(D_\mathbf{x}\times D_\mathbf{v}\times D_u))^n} \leq C < \infty \qquad (7.61)$$

for some $p > 2$, *and such that each* f_i^ε *converges pointwise. We also assume that the microscopic state space has finite measure and that the probability densities* $\mathcal{B}_{jk}$, $\mathcal{C}_{jk}$, $\mathcal{B}_{jk}^e$ *and* $\mathcal{B}_{jk}^e$ *are bounded functions while the interactions rates* η_{ij} *and* η_{ij}^e, *and proliferation/destruction rates* μ_{hk}^i *are all square integrable with respect to their variables. Finally, we assume that* $\mu_{i,j}^1$ *and* $\mu_{i,j}^2$ *are continuous. Then, the pointwise limit of* $\mathbf{f}^\varepsilon$ *is the vector valued function* $\mathbf{M} = \{M_{\rho_i^0,\mathbf{U}_i^0}^i\}_{i=1}^n$ *given by Assumption 7.8 with*

$$\rho_i^0 = \lim_{\varepsilon \to 0} \rho[f_i^\varepsilon], \qquad \mathbf{U}_i^0 = \lim_{\varepsilon \to 0} \mathbf{U}_i^\varepsilon,$$

i.e., the weak and pointwise limits of the local density and the local velocity of f_ε. *Moreover, in the three regimes introduced above, the limiting densities* ρ_i^0 *and velocities* $\mathbf{U}_i^0$ *verify (7.58), (7.59) and (7.60), respectively.*

Remark 7.3. The number of functional subsystem may be variable in time from the initial number n_0 at the initial time t_0. This is due to the onset of new subsystems generated by mutations.

Remark 7.4. The same analysis can also be developed in the case of the parabolic scaling. An example is given in [Bellomo, Bellouquid, and Herrero (2007)] for a closed system in the absence of mutations. The analysis can be technically generalized to the more general case considered in this section.

7.5   On the Molecular-Cellular Scale Modeling

It has been shown in the preceding subsection how macroscopic models can be related to the underlying models at the cellular scale. The mathematical structures at the macroscopic scale depend on the biological rates expressed by cells. These rates are, in each model, free parameters to be adjusted to empirical data. On the other hand, it is well understood that cellular properties depend on the dynamics at the molecular (genetic) scale, as it is documented in the books [Frank (2007)] and [Weinberg (2007)].

The research activity in the field is addressed to the identifications, by microarrays and other technical devices, to the identification of genes whose higher and lower expressions can induce specific pathologies. The literature on this type of investigation is rapidly growing as documented in [Arnsdorf, Tummala, and Jacobs (2009); Hay (2009); Kuemer, Tacheuchi, and Quinlan (2001); Mani *et al.* (2007); Yang (2004)] and many other papers published in this century.

However, despite the afore-said vast literature, the main problem consists in studying the dynamics at the molecular scale. Although contributions in this field is still at an early stage, some interesting results concerning the interpretation of cell dynamics and mutation as an adaptive evolutionary process are available. Relevant contributions have been given by Nowak *et al.* [Nowak and Sigmund (2004)] whose theory has been put in a mathematical framework by a school of researchers. In particular, by Komarova *et al.* [Komarova and Wodarz (2004); Komarova (2006, 2007)], who classified the various loss and gain events of gene expression in the adaptive evolution assigning to each of them a suitable probability related to different stages of the evolution. Detailed analysis focused on referring the afore mentioned gain and loss events to shape compatibility have been proposed in [Gabetta and Ragazzini (2010); Tiuryn, Wójtowicz, and Rudnicky (2007)]. In general, models need to consider the adaptive and evolutionary aspects of gene expression [Tannenbaum and Shakhnovich (2005)].

The modeling approach can take advantage of the reasoning proposed in [Bellomo and Forni (2008)], where it is observed that the cellular dynamics is related to the expression of genes, that is heterogeneously distributed. Some perspective ideas are proposed in [Bellomo and Delitala (2008)] still wait to be developed within a proper research program.

Bearing all the above in mind, let us summarize some preliminary ideas that can possibly be developed towards the ambitious objective of linking the molecular scale to the cellular one. A conceivable strategy consists in considering the genes somehow ordered and in computing the expression:

$$\varphi = \varphi(v) : D_v \to \mathbb{R}_+, \quad D_v \subset \mathbb{R},$$

where the variable v represents the expression referred to the standard one corresponding to the phenotype. Subsequently, a weighted metrics should be created to identify the over or low-expression of certain genes focusing on the onset of anomalous behaviors. For instance:

$$d_p = \int_{D_v} w_p(v)\,\varphi(v)\,dv, \tag{7.62}$$

where d_p can be a positive or negative quantity and w_p is a weight function, which localize the genes whose over- or lower-expression can possibly generate a pathology. More in general, the quantities v, η, and μ can be related to d_p. Specifically to d_v, d_η, and d_μ.

In real biological conditions, the steady description should be replaced by a dynamical system corresponding to functional subsystems of genes and cells respectively, corresponding to the molecular (genetic) scale and the cellular scale.

The overall state of the system, at the higher and lower scales, respectively, is defined by the probability distributions:

$$f = f(t,u) : [0,T] \times D_u \rightarrow \mathbb{R}_+,$$

and

$$\varphi = \varphi(t,v) : [0,T] \times D_v \rightarrow \mathbb{R}_+,$$

over the microscopic states $u \in D_u$ and $v \in D_v$ of the interacting entities regarded as active particles.

The interaction scheme from the lower to the higher scale can be modeled by the following dynamics:

• The evolution of the system at the lower scale is determined by the interaction between genes among themselves and with the outer environment that is supposed to be known and represented by a term $\psi = \psi(t,v)$.

• The evolution of the system at the higher scale is determined by the interaction between active particles, of the population, among themselves and with particles of the lower system that is obtained by solution of the evolution equation for such a system.

Methods of the mathematical kinetic theory of active particles reviewed in Chapter 6 can be used to put the above formal scheme into mathematical structure. Technical calculations yield:

$$\begin{cases} \partial_t \varphi(t,v) = J_1[\varphi,\varphi](t,v) + Q_1[\varphi,\psi](t,v), \\ \\ \partial_t f(t,u) = J_2[f,f](t,u,d_p(t)) + Q_2[f,\varphi](t,u), \end{cases} \tag{7.63}$$

where $J_1[\varphi,\varphi]$ takes into account the interaction within active particles at the lower scale; $Q_1[\varphi,\psi]$ their interactions with the outer environment; $J_2[f,f]$ takes into account the interaction within active particles at the higher scale. This coupling action is not autonomous.

The above structure can be regarded as a candidate to model the dynamical gene expression and their influence over cellular dynamics for minimal models. A necessary generalization consists in identifying for both systems at a different scale the functional subsystems that apply a certain function, namely generation of phenotypes from the gene-expressions and cellular functions generated by the terms d_p^i corresponding to the i-th functional subsystem at the molecular level.

7.6 Critical Analysis

The contents of this chapter has been devoted to showing how macroscopic models of biological tissues can be obtained by a suitable asymptotic analysis applied to the equations modeling the underlying description delivered by the kinetic theory for active particles. Subsequently, some perspective ideas have been given to deal with the link from the molecular to the cellular scale. This specific challenging problem has not yet received a satisfactory answer. On the other hand, it focus the necessary background needed for the derivation of a mathematical theory of complex biological systems.

Referring to the first part of this volume, namely the link between cells and tissues, it has been shown [Bellomo *et al.* (2010)], how the hyperbolic equations can be derived corresponding to different rates of mechanical encounters, proliferative and/or destructive rates, and mutations. These equations include the modeling of macroscopic equations generated by therapeutical actions delivered at the cellular scale. Interactions with the lower scale shows that the characteristics of tissues evolve in time and may even change of type. The asymptotic analysis has been offered as a mathematical tool to be applied to specific biophysical phenomena. In particular, the derivation of chemotactic models and flux limited models is dealt with in [Bellomo *et al.* (2010)].

More in general, the tools offered by this chapter, as well as the preceding one, provides a general framework to derive specific models. The application presented in Chapter 9 will show the use of these tools to derive a mathematical model of interest for the applications. All models are characterized by parameters that have to be implemented on the basis of empirical data. In other words, the model needs to be related to a specific experiment and, if validated, can be used to reproduce biological events somewhat different from those of the experiment.

One of the various conceivable uses of models is the parameter sensitivity analysis, which aims at showing emerging behaviors that appear for particular values, or range of values, of the parameters. An example of analysis of this type is given in paper [De Angelis *et al.* (2003)], where two different models of competition between cancer and immune cells have been examined. Specifically, it has been shown that a parameter related to the ability of the immune system to learn the presence of cancer cells has a crucial role on the output of the competition between immune system and cancer. Namely, there exists a bifurcation value of such a parameter such that for values higher than it, the immune system prevails with progressive destruction of immune cells, while for lower values the number of cancer cells grows with progressive inhibition of the immune system. An important application is the simulation of therapeutical actions that may be technically related to the above sensitivity analysis with the aim of optimizing the application of specific therapies. The literature in this field is rapidly growing as documented in [Bellomo *et al.* (2010); Frank, Herty, and Schäfer (2008)] and many others.

However, the above apparently optimistic description can be misleading unless all conceptual difficulties are properly analyzed. The main problem, in the authors' opinion, is the selection of the modeling scale and, subsequently, the study of the interactions between the lower and higher scales. This is a crucial step in the development of a mathematical-biological theory as already anticipated in the first chapter. This chapter, which concludes the introduction to mathematical tools, has selected the representation at the cellular scale and deal with the mathematical approach to model the links with the lower and higher scales of genes and tissues respectively.

Moreover, the analysis has included the role of the immune system that may contribute to contrast pathological states, but may also negatively contribute to a Darwinian-type selection of cells carriers of aggressive undesired states. Hopefully, the approach should also contribute to the validation of models by depicting emerging behaviors that are typical of complex systems, and hence of those treated in these Lectures Notes. As a matter of fact, the identification of parameters based on empirical data is often limited to the conditions of the experiment and fails when these conditions are modified.

Finally, it is worth stressing that the contents of this chapter also stimulates research activity of applied mathematicians motivated by a variety of analytic problems, which are still open. For instance dealing with the derivation of macroscopic equations when the terms η and μ depend on the state of the interacting cells or even on the distribution function. This situation rises in the case of nonlinear interactions.

Although the asymptotic analysis has been focused on the hyperbolic scaling, also the derivation of diffusion models is of interest for the applications. Particularly interesting is the case of the derivation of flux limited models that appear to be closer to physical reality with respect to the classical parabolic models. The two-scale approach, that has been mentioned above, may possibly lead to new models beyond the heuristic derivation based on physical intuitions.

PART 3
Applications and Research Perspectives

Chapter 8

A Model for Malign Keloid Formation and Immune System Competition

8.1 Introduction

This chapter presents a mathematical model related to the competition among viruses, keloid cells, and malignant cells, on one hand and immune system cells on the other. The model is developed in the framework of the kinetic theory of active particles proposed in Chapter 6. Mathematical models can be designed by identifying the functional subsystems involved in the process, each of them characterized by a certain biological activity, and, according to the biological phenomena under consideration, by modeling the interactions at the microscopic level between cells of the same or of different subsystems. The predictive capability of the model is tested by means of numerical simulations which show the typical emerging phenomena of the disease.

The sequential phases of Keloid formation and related wound healing process were described in Part I. As already mentioned, the failure of one of these phases may generate fibrotic diseases. This section deals with the phenomenological description, in view of the modeling approach, of keloid formation and with the medical hypothesis that triggers the formation process.

Keloid is a benign dermal tumor that forms during a protracted wound healing process [English and Shenefelt (1999); Marneros *et al.* (2004); Niessen *et al.* (1999)]. It is characterized by an increased deposition of the extracellular matrix by mutated fibroblast cells. After deposition of collagen, a hard, thick lump of overgrown tissue develops as the result of insufficient resorption and apoptosis. The original form is not optimally restored. The abnormal form of the scar remains due to a hardening tendency.

The triggering causes and the key alterations responsible for keloid formation remain elusive and there is no satisfactory treatment for this disorder. Recently it has been proposed an infection-based hypothesis (specifically, a viral hypothesis) [Alonso, Rioja, and Pera (2008)]. Specifically, it has been hypoth-esized that healthy individuals carrier of a virus, either known or unknown, as-sociated to some adjuvant, and characterized by some genetic susceptibility, may develop keloid during the scar maturation process. The virus uses the bone mar-row or lymphatic system as its reservoir, residing there in a silent state, and reaches the wound via an internal circuit through which the viral genome is transported from its myeloid reservoir to the wound via bone marrow or circulating fibrocytes chemotactically attracted to the damaged skin region. Once in the wound, the virus can switch from a silent to a latent state by effect of some chemical stimu-lus that is probably generated during the tissue repair process. In the new state, the transcription of some of the powerful viral proteins might cause thorough de-railment of the normal repair process. As a result, keloid growth might depend on both individual susceptibility and on the viral load deposited in the wound; the greater the susceptibility and viral load are, the more markedly the keloid develops into an aggressive state.

Focusing on growth potential, heterogeneous fibroblast cells [Placid and Lewis (1992)], metabolic differences and cell degeneration/apoptosis are present in different areas of keloids. According to [Naitoh *et al.* (2005); Saed *et al.* (1998)], these fibroblast cells eventually undergo a somatic mutation in tumor sup-pressor p53, and in addition to the inherited predisposition, generate cells (keloid) which have a high potential proliferation rate and an ability to escape apopto-sis. There is evidence that keloid fibroblast cells have key pathways that are al-tered compared with normal fibroblast cells, during a number of stress responses [Nakaoka, Miyauchi, and Miki (1995)]. This implies that the keloid might also contribute to the formation of malignancies (neoplasia).

It is worth pointing out that there is a higher occurrence of keloid in darker-skinned races and those of Asian descent, but a difference in occurrence based on sex has not been demonstrated convincingly. Moreover, the penetrance of keloid formation is age related (keloid usually occurs between age 10 and 30 years of age, at which time plasma levels of the growth hormone and insulin-like growth factor 1 are also high).

As already mentioned, one way of overcoming the complexity arising from the mathematical representation of a system at the microscopic scale can be found in the statistical description of the system, applying the tools of the kinetic theory for active particles. Accordingly, the system is described by a set of distribution functions over the microscopic state of the interacting functional subsystems. The derivation of the equations that describe the evolution of the aforementioned dis-tribution is obtained for conservation equations in the space of the microscopic states, while the fluxes of the particles of the functional subsystems are computed by taking advantage of microscopic interaction models.

The mathematical model we present in this chapter concerns the competition among viruses, keloid cells, and malignant cells, on one hand and immune system cells on the other. The model is developed taking into account the above mentioned medical hypotheses which state that viruses and the genetic susceptibility of patients are the main causes that trigger the formation of a keloid and its malignant effects. Obviously, the final target is to tune of the model according to data on the patients provided by physicians.

The contents of this chapter are proposed in three more sections, which follow this introduction. Section 8.3 deals with the derivation of a mathematical model that is suitable for describing the evolution and competition among functional subsystems. Section 8.4 develops some numerical simulations, which show the typical emerging phenomena of keloid formation and its possible malignant effects. Finally, Section 8.5 concerns a critical analysis of the contents of this chapter, in view of possible further refinements, which should be regarded as a research perspective.

8.2 The Mathematical Model

The mathematical model proposed in this section refers to [C. Bianca (2010)] and is derived at the cellular scale according to the phenomenological description reported in Part I. As already mentioned, the modeling is developed by using the general mathematical framework of the kinetic theory for active particles presented in Chapter 6. Therefore, the approach differs from that at the macroscopic scale.

According to the KTAP theory, first we deal with the characterization of the functional subsystems and their statistical representation, subsequently with the modeling of the microscopic interactions between the selected functional subsystems, and finally with the derivation of the relative class of evolution equations.

Bearing all the above in mind, the following assumptions are proposed:

Assumption 8.1. (Functional Subsystems) The keloid formation process involves four functional subsystems that interact with each other. Moreover, their interaction may result in the genesis of a new functional subsystem constituted by cells with a high degree of malignancy, e.g. tumor cells. Specifically the following interacting functional subsystems are involved:

(1) *Normal-Fibroblast cells* (NFc). These cells are the most common cells of connective tissue and are able to synthesize the extracellular matrix and colla-

gen, the structural framework (stroma) for animal tissues. Fibroblast cells are morphologically heterogeneous with diverse appearances depending on their location and biological function.

(2) *Activated-Viruses* (AV). Viruses are infectious agents that can replicate only inside the cells, which, as consequence of the infections, became viral cells or activated-viruses.

(3) *Keloid-Fibroblast cells* (KFc). A normal fibroblast cell may undergo a mutation as consequence of a viral action or genetic susceptibility. This mutation gives the cells a significant advantage with regard to proliferation, the possibility to escape apoptosis, and allows it and its descendant to quickly advance along the keloid.

(4) *Malignant cells* (Mc). Activated viruses may damage a keloid fibroblast cell and generate a malignant cell.

(5) *Immune system cells* (ISc). These cells have the ability to recognize and contrast KFc, Mc, and AV through recognition and destruction.

Assumption 8.2. (Activity State) Each functional subsystem described in the previous assumption is able to express a well-defined biological function represented by the scalar variable $u \in D_u = [0, \infty[$. Specifically:

(1) NFc: The activity variable refers to the *proliferation ability*. Increasing values of u indicate an increasing intensity of proliferation.

(2) AV: The activity variable u is a magnitude of their *aggressiveness* related to their proliferation ability.

(3) KFc: The activity variable u corresponds to the *proliferation ability* but these cells are characterized by a relatively greater ability to proliferate with respect to the normal fibroblast cells (*genetic instability*).

(4) Mc: The activity variable is a magnitude of their progression ability. These cells are characterized by a greater and greater genetic instability with respect to the KFc. Increasing values of u take into account these capabilities.

(5) ISc: The activity variable refers to the degree of activation and of the response

to foreign agents.

It is worth mentioning that the immune system is constituted by several different cell types; however to simplify the complexity induced by considering a large number of subsystems, we will consider the immune system as one only functional subsystem. Therefore it develops activities which actually are heterogeneously distributed among several particular subsystems.

In the sequel, when we refer to subsystems 2, 3, and 4 as a whole, we will call them **non-self cells**.

Referring to the mathematical framework proposed in Chapter 6 and to the above defined subsystems, the derivation of evolution equations for the distribution functions of each subsystem needs the modeling of the microscopic interactions among individuals of the various subsystems.

In the sequel, we will discuss only conservative, proliferative, and transition interaction terms. Interactions which have nontrivial outputs, i.e., an effective change (either in the microscopic state or in the entities number) occurs after the interaction, are not considered. Thus biological phenomena corresponding to interactions that give trivial outputs will not be taken into account, as for instance the physiological birth and death of the entities.

The microscopic interaction terms are derived under the following assumptions:

Assumption 8.3. (Statistical representation) The distribution of the functional subsystems is described by the functions $f_1(t,u)$ (normal fibroblast cells), $f_2(t,u)$ (activated viruses), $f_3(t,u)$ (keloid fibroblast cells), $f_4(t,u)$ (malignant cells), and $f_5(t,u)$ (immune system cells).

We are interested in modeling the keloid formation in a specific body part (hand, leg, face, etc.), while the following assumptions are introduced toward the derivation of the model:

Assumption 8.4. The system is assumed to be homogeneously distributed in space. According to a mean field approximation, the encounter rate is assumed to be constant for all interacting pairs; for simplicity, $\eta_{ij} = 1$ for all $i, j \in \{1,2,3,4,5\}$.

Assumption 8.5. The term $\mathscr{B}_{ij}$ modeling the transition probability density is assumed, for $i, j \in \{1,2,3,4,5\}$, to be defined by a delta Dirac function with the most probable output m_{ij} depending on the microscopic state of the interacting pairs:

$$\mathscr{B}_{ij}(u_* \rightarrow u \,|\, u_*, u^*) = \delta(u - m_{ij}(u_*, u^*)). \tag{8.1}$$

According to assumptions, the modeling of each term of interactions is obtained as follows:

• **Conservative interactions.**
It is assumed that AV, KFc, and Mc are the only subsystems subject to conservative interactions, therefore:
$$C_1[\mathbf{f}](t,u) = C_5[\mathbf{f}](t,u) = 0.$$
Moreover, the conservative terms $C_2[\mathbf{f}](t,u)$, $C_3[\mathbf{f}](t,u)$, and $C_4[\mathbf{f}](t,u)$ are derived under the assumptions that the non-self cells have a tendency to increase their microscopic state with a certain rate, regulated by the interactions with the NFc and AV. The evolution toward higher level of activity is called **self-proliferation** for KFc and AV instead is called **self-progression** for MC. Moreover, it is assumed that the self-proliferation of KFc is greater than the self-proliferation of AV. Accordingly to Eq. (8.1), we define

$$m_{ij}(u_*, u^*) = \begin{cases} u_* + \varepsilon\alpha & \text{if } j = 1 \text{ and } i = 2, \\[2mm] u_* + \alpha & \text{if } j = 2 \text{ and } i = 3, \\[2mm] u_* + \varepsilon^2\alpha & \text{if } j = 2 \text{ and } i = 4, \\[2mm] u_* & \text{otherwise,} \end{cases} \tag{8.2}$$

where α is a positive parameter, which defines the self-proliferation rate toward high states of activity for KFc, while $\varepsilon < 1$ is a scale parameter, which takes into account the difference among the self-proliferation and the self-progression rates of the AV and Mc with respect to the KFc.

Straightforward calculations yield:

$$C_i[\mathbf{f}](t,u) = \begin{cases} n[f_1](t)\,[f_2(t, u - \varepsilon\,\alpha) - f_2(t,u)] & \text{if } i = 2, \\[2mm] n[f_2](t)\,[f_3(t, u - \alpha) - f_3(t,u)] & \text{if } i = 3, \\[2mm] n[f_2](t)[f_4(t, u - \varepsilon^2\alpha) - f_4(t,u)] & \text{if } i = 4, \\[2mm] 0 & \text{otherwise.} \end{cases} \tag{8.3}$$

• **Proliferative interactions without Transition of Subsystem.**
Each subsystem can proliferate (without transition of subsystem) when it encounters another subsystem. Specifically:

P.1. NFc proliferate when encounter each other;

P.2. AV proliferate when encounter NFc and ISc;

P.3. KFc proliferate when encounter NFc and AV;

P.4. Mc proliferate when encounter AV;

P.5. ISc proliferate when encounter AV, KFc, and Mc.

Proliferation rates

It is assumed that the proliferation rate of KFc cell is greater than the proliferation rates of NFc, AV, and Mc; the proliferation rate of the ISc, when encounter AV and Mc, is greater than the proliferation rate of the ISc when encounter KFc. Accordingly, we define:

$$
\mu_{ij}(u,u^*) = \begin{cases} \varepsilon^2\beta & \text{if } j = 1 \text{ and } i = 1, \\[1.5ex] \varepsilon\beta & \text{if } j \in \{1,5\} \text{ and } i = 2, \\[1.5ex] \beta & \text{if } j \in \{1,2\} \text{ and } i = 3, \\[1.5ex] \varepsilon\beta & \text{if } j = 2 \text{ and } i = 4, \\[1.5ex] \beta_I & \text{if } j \in \{2,4\} \text{ and } i = 5, \\[1.5ex] \varepsilon^2\beta_I & \text{if } j = 3 \text{ and } i = 5, \\[1.5ex] 0 & \text{otherwise,} \end{cases} \tag{8.4}
$$

where β is a positive parameter corresponding to the proliferation rate of KFc; β_I is the proliferation rate of the ISc, and ε is the scale parameter. The proliferation terms thus read:

$$P_i[\mathbf{f}](t,u) = \begin{cases} \varepsilon^2 \beta f_1(t,u)\, n[f_1](t) & \text{if } i = 1, \\[2ex] \varepsilon \beta f_2(t,u)\, [n[f_1](t) + n[f_5](t)] & \text{if } i = 2, \\[2ex] \beta f_3(t,u)\, [n[f_1](t) + n[f_2](t)] & \text{if } i = 3, \\[2ex] \varepsilon \beta f_4(t,u)\, n[f_2](t) & \text{if } i = 4, \\[2ex] \beta_I f_5(t,u)\, [n[f_2](t) + \varepsilon^2 n[f_3](t) + n[f_4](t)] & \text{if } i = 5. \end{cases} \tag{8.5}$$

• **Destructive interactions without Transition of Subsystem.**
The cells of each subsystem may be destroyed when they encounter cells of another subsystem. Specifically:

D.1. NFc are destroyed by AV;

D.2. AV and Mc are destroyed by ISc;

D.3. KFc are destroyed by AV and ISc;

D.4. ISc are destroyed by AV, KFc, and Mc.

Destruction rates
ISc are able to destroy (with rate δ) AV and Mc more efficiently than the KFc; the latter are destroyed by AV less efficiently than NFc.

The non-self cells with a high level of activity have the ability to inhibit or destroy immune system cells (*immune suppression* or *immune-subversion*) but the destruction rate of the ISc by KFc is assumed to be less than the destruction rate by Mc and AV. It is worth stressing that we assume a destruction of the immune cells by non-self entities, since inhibited immune cells do not play a relevant role in the competition and may be equivalently assumed as eliminated. Accordingly, we define:

$$
\mu_{ij}(u,u^*) = \begin{cases}
-\varepsilon\delta & \text{if } j = 2 \text{ and } i = 1, \\[2mm]
-\delta & \text{if } j = 5 \text{ and } i \in \{2,4\}, \\[2mm]
-\varepsilon^2\delta & \text{if } j \in \{2,5\} \text{ and } i = 3, \\[2mm]
-\varepsilon^2\delta_i u^* & \text{if } j = 3 \text{ and } i = 5, \\[2mm]
-\delta_I u^* & \text{if } j \in \{2,4\} \text{ and } i = 5, \\[2mm]
0 & \text{otherwise,}
\end{cases}
\tag{8.6}
$$

where δ_I is a positive parameter which characterizes the destruction of the immune system cells by AV and Mc. Thus the destructive terms read:

$$
D_i[\mathbf{f}](t,u) = \begin{cases}
-\varepsilon\delta f_1(t,u)\,n[f_2](t) & \text{if } i = 1, \\[2mm]
-\delta f_2(t,u)\,n[f_5](t) & \text{if } i = 2, \\[2mm]
-\varepsilon^2\delta f_3(t,u)\,[n[f_2](t) + n[f_5](t)] & \text{if } i = 3, \\[2mm]
-\delta f_4(t,u)\,n[f_5](t) & \text{if } i = 4, \\[2mm]
-\delta_I f_5(t,u)\,[a[f_2](t) + \varepsilon^2 a[f_3](t) + a[f_4](t)] & \text{if } i = 5.
\end{cases}
\tag{8.7}
$$

- **Interactions with Transition of Subsystem.**

It is assumed that cell interactions may occur which result, due also to genetic mutations, to birth of a new subsystem. Specifically:

T.1. NFc may pass in KFc when interact with NFc;

T.2. NFc may pass in KFc when interact with AV;

T.3. KFc may pass in Mc when interact with AV.

In particular, we assume that it is more likely that NFc become KFc when they encounter AV. Moreover, the microscopic state of the entities does not change during the transition. Accordingly, we define

$$\mu_{hk}^i(u_*, u^*; u) = \begin{cases} \varepsilon\gamma\delta(u - u_*) & \text{if } h = 1, k = 1, \text{ and } i = 3, \\ \gamma\delta(u - u_*) & \text{if } h = 1, k = 2, \text{ and } i = 3, \\ \lambda\,\delta(u - u_*) & \text{if } h = 3, k = 2, \text{ and } i = 4, \\ 0 & \text{otherwise,} \end{cases} \tag{8.8}$$

where γ and λ are the transition rates in the KFc subsystem and in the Mc subsystem, respectively. Thus the transition terms read:

$$T_i[\mathbf{f}](t, u) = \begin{cases} \gamma f_1(t, u)\left[\varepsilon\, n[f_1](t) + n[f_2](t)\right] & \text{if } i = 3, \\ \lambda f_3(t, u)\, n[f_2](t) & \text{if } i = 4, \\ 0 & \text{otherwise.} \end{cases} \tag{8.9}$$

• **The Evolution Equations.** The mathematical model consists in an evolution equation for the distribution functions f_i, for $i \in \{1, 2, 3, 4, 5\}$, corresponding to the previous mentioned subsystems. Based on the previous modeling of subsystem interactions, we are now able to derive the evolution equations by replacing the previous terms into the general framework described in Chapter 6. Straightforward calculations yield:

$$
\begin{cases}
\partial_t f_1 = \varepsilon \left(\varepsilon \beta \int_0^\infty f_1(t,u)\,du - \delta \int_0^\infty f_2(t,u)\,du \right) f_1(t,u), \\[2ex]
\partial_t f_2 = \left(\varepsilon \beta \int_0^\infty [f_1(t,u) + f_5(t,u)]\,du - \delta \int_0^\infty f_5(t,u)\,du \right) f_2(t,u) \\[2ex]
\qquad - f_2(t,u) \int_0^\infty f_1(t,u)\,du + f_2(t, u - \varepsilon\alpha) \int_0^\infty f_1(t,u)\,du, \\[2ex]
\partial_t f_3 = \left(\beta \int_0^\infty f_1(t,u)\,du - (1 - \beta + \varepsilon^2 \delta) \int_0^\infty f_2(t,u)\,du \right) f_3(t,u) \\[2ex]
\qquad - \varepsilon^2 \delta\, f_3(t,u) \int_0^\infty f_5(t,u)\,du + f_3(t, u - \alpha) \int_0^\infty f_2(t,u)\,du, \\[2ex]
\partial_t f_4 = \left((\varepsilon \beta - 1) \int_0^\infty f_2(t,u)\,du - \delta \int_0^\infty f_5(t,u)\,du \right) f_4(t,u) \\[2ex]
\qquad + \lambda\, f_3(t,u) \int_0^\infty f_2(t,u)\,du + f_4(t, u - \varepsilon^2 \alpha) \int_0^\infty f_2(t,u)\,du, \\[2ex]
\partial_t f_5 = \beta_I \left(\int_0^\infty [f_2(t,u) + f_4(t,u)]\,du + \varepsilon^2 \int_0^\infty f_3(t,u)\,du \right) f_5(t,u) \\[2ex]
\qquad - \delta_I \left(\int_0^\infty u\,[f_2(t,u) + f_4(t,u)]\,du + \varepsilon^2 \int_0^\infty u\, f_3(t,u)\,du \right) f_5(t,u).
\end{cases}
\tag{8.10}
$$

Model (8.10) consists in a system of nonlinear integro-differential equations, with quadratic type nonlinearity, which is characterized by 8 phenomenological parameters having a well-defined biological meaning. Their detailed description is given in Table 8.1 where functional subsystems, relevant interaction phenomena, and related parameters are summarized. All parameters have to be regarded as positive constants (eventually equal to zero), small with respect to unity, and have to be identified by suitable experiments.

The parameters can be classified into four groups:

The α-parameter that refers to mass conservative encounters;

the β-parameters that refer to encounters which generate proliferative events;

the δ parameters that refer to destructive interactions, the parameters γ and λ are related to transition encounters;

while ε is a scale parameter.

A detailed description is given in Table 8.1 where functional subsystems, relevant interaction phenomena, and related parameters are summarized. All parameters have to be regarded as positive constants (eventually equal to zero), small with respect to unity, and have to be identified by suitable experiments.

Table 8.1 Functional subsystems, interaction terms, and related parameters.

	INTERACTIONS	(NFc)	(AV)	(KFc)	(Mc)	(ISc)
(NFc)	Proliferative	$\varepsilon^2\beta$				
	Destructive		$-\varepsilon\delta$			
	Transitive	$\varepsilon\gamma$	γ			
(AV)	Conservative	$\varepsilon\alpha$				
	Proliferative	$\varepsilon\beta$				$\varepsilon\beta$
	Destructive					$-\delta$
(KFc)	Conservative		α			
	Proliferative	β	β			
	Destructive		$-\varepsilon^2\delta$			$-\varepsilon^2\delta$
	Transitive		λ			
(Mc)	Conservative		$\varepsilon^2\alpha$			
	Proliferative		$\varepsilon\beta$			
	Destructive					$-\delta$
(ISc)	Proliferative		β_I	$\varepsilon^2\beta_I$	β_I	
	Destructive		δ_I	$-\varepsilon^2\delta_I$	δ_I	

8.3   Simulations and Emerging Behaviors

This section deals with simulations of the densities and of the distribution functions of the functional subsystems. It is worth stressing that the simulations developed in this section followed by their medical interpretations do not cover the whole panorama related to the sensitivity analysis of all parameters. However, various aspects of competition that is of interest for the medical sciences are treated.

Simulations are obtained by solving the following initial value problem, for short (IVP):

$$\begin{cases} \partial_t \mathbf{f}(t,u) = \mathbf{J}[\mathbf{f}](t,u)\,, \\[2mm] \mathbf{f}(t=0,u) = \mathbf{f}^0(u)\,, \end{cases} \qquad (8.11)$$

where $\mathbf{J}[\mathbf{f}](t,u)$ represents the right-hand side terms of the system of Eq. (8.10), and $\mathbf{f}^0 = (f_1^0, f_2^0, f_3^0, f_4^0, f_5^0)$ is the vector of the initial conditions.

The proof that the initial value problem (8.11) is well posed locally in time is not reported here since this proof, which is based on classical fixed point arguments, is available in the literature (see [Arlotti, Bellomo, and De Angelis (2002)]). It is worth mentioning that the proof of the local existence and uniqueness of the solution of the (IVP) (8.11) allows us to develop appropriate computational methods to obtain simulations of the model. The computational scheme is that of the well-known generalized collocation method.

Accordingly, the variable u is discretized into a suitable set of collocation points and the distributions functions are interpolated by Sinc functions. Then the integral terms are approximated by means of algebraic weighted sums in the nodal points of the discretization. The particularization of the evolution equations in each node and the enforcing of the initial conditions transform the integro-differential (IVP) into an (IVP) for systems of ordinary differential equations, describing the evolution of the values of the distribution functions in the node of the collocation.

Simulations are technically addressed to show the typical phenomena of the keloid formation and its possible negative effects such as the onset of malignant cells, the heterogeneity of the non-self entities, and the competition among non-self cells and cells of the immune system when varying the magnitude of the parameters.

In particular, as objectives of the computational analysis, we focus on the sensitivity analysis of the following parameters:

- *the progression rate α of the KFc,*

- *the proliferation rate β_I of the ISc.*

Specifically, we assume that seven of the eight parameters of the model are fixed, while one of them, the free parameter, spans from zero to higher values.

The simulations of the above cases start when the number of NFc in the wound is equal to the number of AV, namely

$$n[f_1](0) = n[f_2](0),$$

and a number $n[f_5](0)$ of immune system cells have reached the wound (sentinel level). Thus we consider non-zero initial conditions only for $f_1(0,u)$, $f_2(0,u)$, and $f_5(0,u)$, which is coherent with the physiological situation, where only NFc, AV, and ISc are initially present. Finally, in the last subsection we consider the sensitivity analysis of the initial distributions.

The parameters kept at constant values for all simulations are chosen in a way that:

The transition from the NFc to KFc is not negligible ($\gamma = 0.4$);

The ability of the immune system cells to destroy the non-self cells is quite low ($\delta = 0.3$);

The non-self cells have an intermediate ability to inhibit the response of the immune system cells ($\delta_I = 0.5$);

The scale factor has an intermediate value ($\varepsilon = 0.5$).

On the contrary, the values of the parameters α, β, β_I, and λ are set case-wise as detailed in the next subsections.

8.3.1   *Sensitivity Analysis of the Progression Rate* α

In this subsection we let α vary from low to higher values. Specifically, after having fixed the parameters γ, δ, δ_I, and ε as previously indicated, we set:

$\beta = 0.4$, (weak KFc proliferation);

$\beta_I = 0,35$ (low proliferation of the ISc);

$\lambda = 0.5$ (the probability that a keloid fibroblast cell becomes a malignant cell is not negligible).

Expected asymptotic behavior. In general, we should expect an increasing amplification of the heterogeneity phenomena of the non-self entities, and correspondingly increase the chances to develop malignant effects.

Simulations for values of $\alpha \in [0, 0.35]$. The model is able to reproduce the onset of AV, KFC, and Mc. As showed in Figure 8.1, the low magnitude of the progression rate never allows the number of Mc to overcome the number of KFc.

Moreover, since the number of non-self cells with high activity is low (see the distribution functions of the AV, KFc, Mc in Figures 8.2, 8.3, 8.4 respectively) the immune system cells are able to quickly deplete cells that may generate malignant effects.

BIOLOGICAL INTERPRETATION. These simulations may represent the failure of the normal wound healing process where, because of the low number of non-self entities with a high level of heterogeneity, the ISc would avoid the formation of keloid and malignant effects.

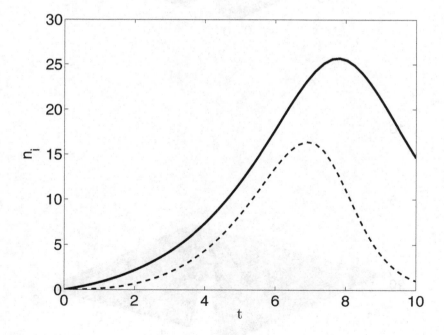

Fig. 8.1   The time evolution of the densities of KFc (solid line) and of Mc (dashed line) for $\alpha = 0.3$. The low magnitude of the progression rate never allows the number of Mc to overcome the number of KFc

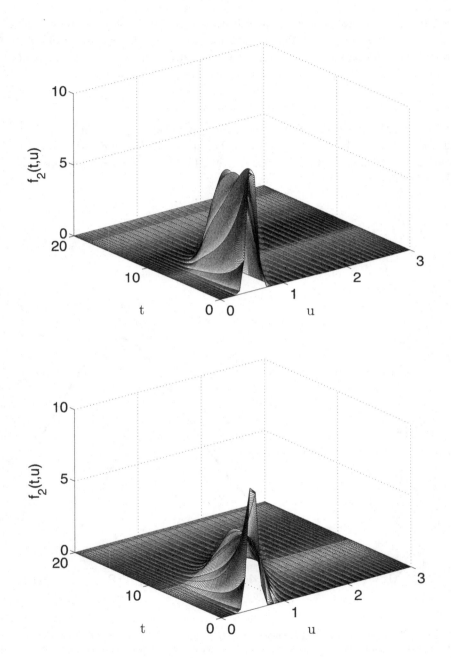

Fig. 8.2   The distribution function of the AV for $\alpha = 0.1$ (top panel) and $\alpha = 0.3$ (bottom panel).

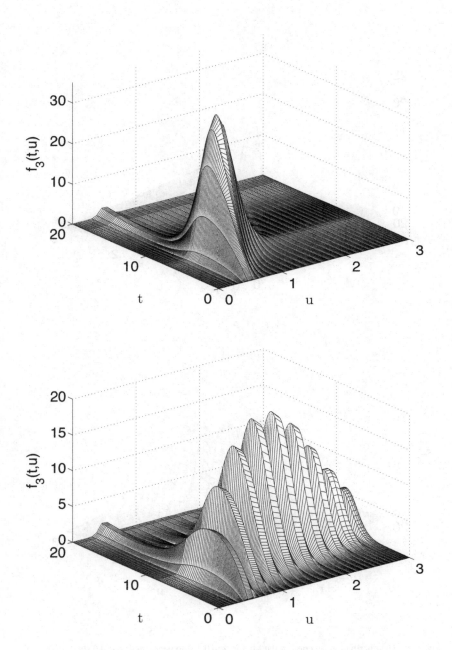

Fig. 8.3   The distribution function of the KFc for $\alpha = 0.1$ (top panel) and $\alpha = 0.3$ (bottom panel).

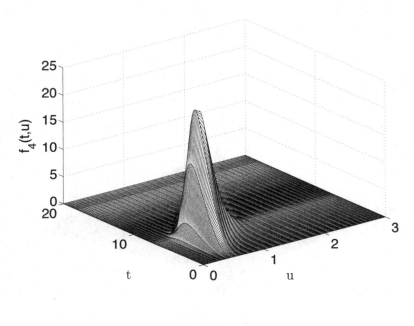

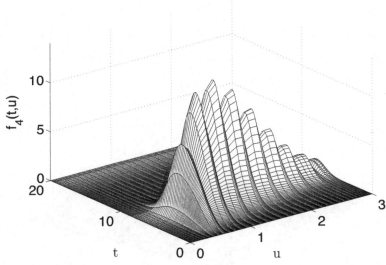

Fig. 8.4   The distribution function of the Mc for $\alpha = 0.1$ (top panel) and $\alpha = 0.3$ (bottom panel).

Simulations for values of $\alpha \in]0.35, 0.5]$. As showed in Figure 8.6, at first stage, the number of KFc is greater than the number of Mc (the latter depends on how many KFc affect the transition of the subsystem) but there is a range where this situation is reversed; the length of this range increases with the magnitude of α. First, we observe that the cells of the immune system must proliferate for a longer period of time before reaching a plateau sufficient to deplete non-self entities (see Figure 8.5). Moreover, increasing values of α produce the birth of non-self entities with high level of activity and consequently more heterogeneity phenomena (compare the distribution functions depicted in Figures 8.7, 8.8, 8.9). Furthermore, the ISc are still able to destroy the non-self entities but the competition among them occurs for a longer period of time with respect to a lower progression rate.

BIOLOGICAL INTERPRETATION. These simulations may represent a failure of the normal wound healing process where the keloid formation, which depends on how long it takes the ISc to deplete the KFc, would prevail malignant effects.

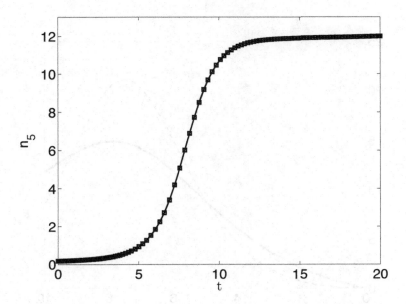

Fig. 8.5   The time evolution of the density of the ISc for $\alpha = 0.5$. The ISc proliferate before reaching a plateau.

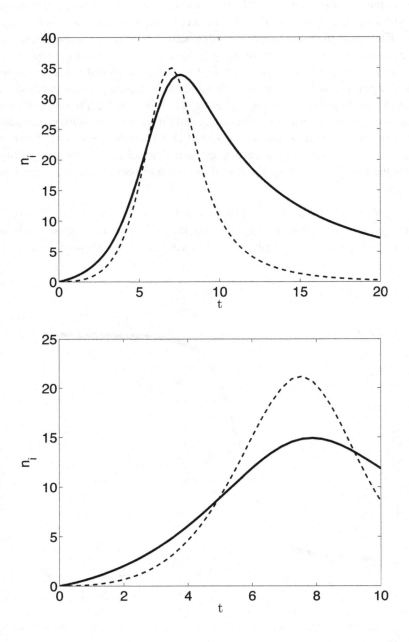

Fig. 8.6   The time evolution of the densities of KFc (solid line) and of Mc (dashed line) for $\alpha = 0.4$ (top panel) and $\alpha = 0.5$ (bottom panel).

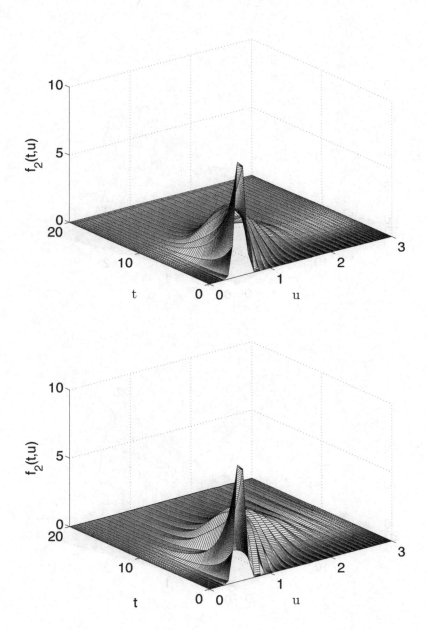

Fig. 8.7 The distribution function of the AV for $\alpha = 0.4$ (top panel) and $\alpha = 0.5$ (bottom panel).

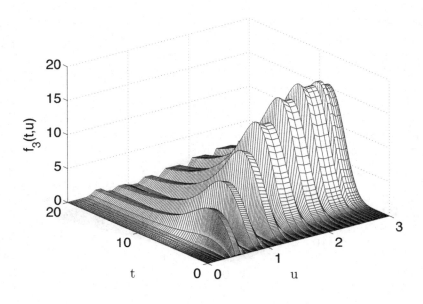

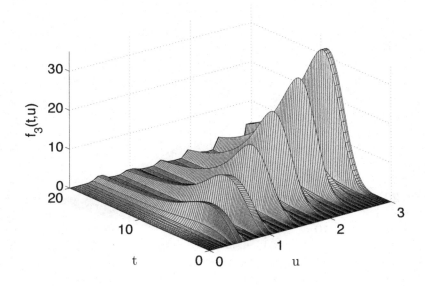

Fig. 8.8   The distribution function of the KFc for $\alpha = 0.4$ (top panel) and $\alpha = 0.5$ (left panel).

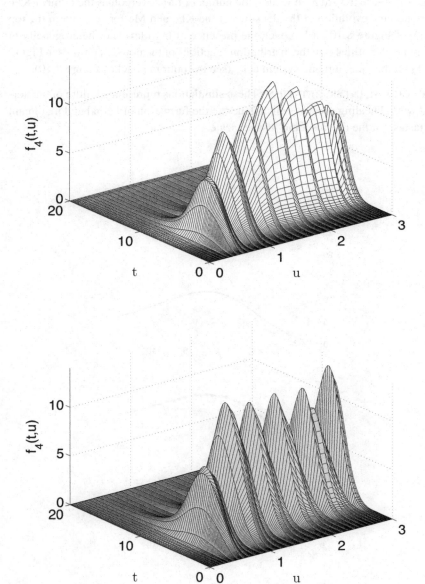

Fig. 8.9 The distribution function of the Mc for $\alpha = 0.4$ (top panel) and $\alpha = 0.5$ (bottom panel).

Simulations for values of $\alpha \in]0.5, 1]$**.** Letting the magnitude of the progression rate α increase toward high values, the number of Mc overcomes the number KFc (see the time evolution of the densities of the KFc and Mc for $\alpha = 0.8$ in the top panel of Figure 8.10). Moreover, the presence of high levels of heterogeneity in the non-self entities (see the distribution functions of the non-self entities in Figure 8.11) inhibits the immune system cells (see the bottom panel of Figure 8.10).

BIOLOGICAL INTERPRETATION. These simulations represent a failure of the normal wound healing process which generate the formation of keloid and malignant metastasis at the macroscopic (tissue) scale.

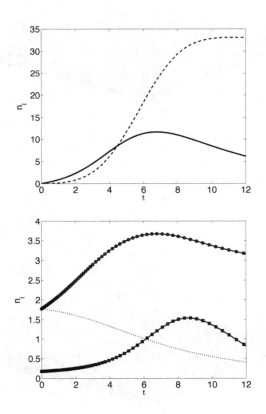

Fig. 8.10   Time evolution of the densities of the KFc (solid line) and the Mc (dashed line) at the top panel and of the NFc (dotted line), AV (circle line), and ISc (square line) at the bottom panel ($\alpha = 0.8$).

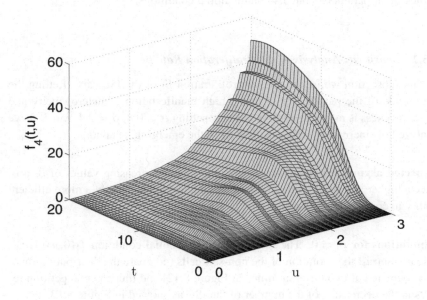

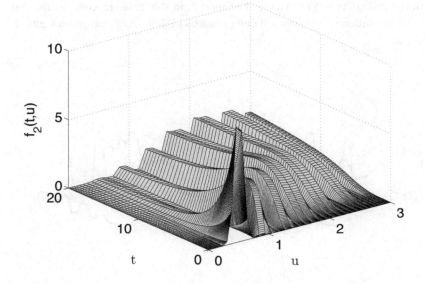

Fig. 8.11 Distribution functions of the Mc (top panel) and AV (bottom panel) for $\alpha = 0.8$.

It is worth stressing that these scenarios depend not only on the specific choice of all parameters but also on the initial conditions.

8.3.2   *Sensitivity Analysis of the Proliferation Rate β_I*

In this subsection we let the rate of proliferation β_I of the ISc vary. Letting the magnitude of the parameters so that a high manifestation of heterogeneity and aggressiveness is manifest, and specifically setting $\alpha = 0.8$, $\beta = 0.4$, $\lambda = 0.5$, we analyze how increasing values of β_I modify the emerging behaviors.

Expected asymptotic behavior. It is expected that increasing values of β_I produce a higher activation of the immune system and consequently a more efficient ability to contrast the non-self cells.

Simulations for $\beta_I = 0$. The sentinel level (the initial condition $f_4(0, u)$) is not able to contrast the evolution of the non-self cells (compare the distribution functions depicted at two different times in Figure 8.12) and thus the competition results in the decreasing of the number of the ISC as showed in Figure 8.13.

BIOLOGICAL INTERPRETATION. The model, in this case, reproduces the progressive evolution of the non-self cells towards high level of malignancy and infections.

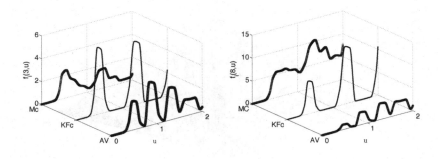

Fig. 8.12   The distribution functions of AV, KFc, and Mc at times $t = 3$ (top panel) and $t = 8$ (bottom panel) for $\beta_I = 0$.

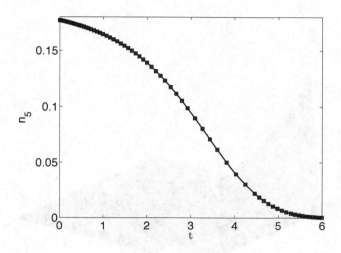

Fig. 8.13   The time evolution of the number density of the ISc for $\beta_I = 0$.

Simulations for $\beta_I \in]0, 0.55]$. There exists an interval of time where the number of Mc is always greater than the number of KFc, namely the immune system does not proliferate enough to avoid that the Mc prevail on the KFc. Moreover, the ISc are able to inhibit partially only the growth of the non-self entities with low levels of activity.

BIOLOGICAL INTERPRETATION. These simulations represent a failure of the normal wound healing process where high levels of proliferation may be captured.

Simulations for $\beta_I \in]0.55, 1]$. The immune system is able to deplete AV and Mc; the efficient of the action increases with the magnitude of β_I. The destruction of the AV and Mc restores the free proliferation of NFc, which, because of the genetic susceptibility, generates the rebirth of KFc (see the distribution functions in Figure 8.14).

BIOLOGICAL INTERPRETATION. The immune system prevents the formation of malignant tumors, but the genetic susceptibility of the patient does not avoid the possibility of the keloid formation.

Further investigations provide that the times from which the number of the non-self cells start to decrease are decreasing function of β_I.

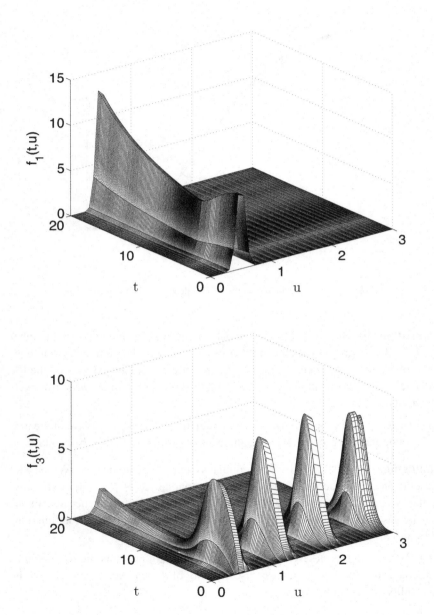

Fig. 8.14   Distribution function of the NFc (top panel) and KFc (bottom panel) for $\beta_I = 0.8$.

8.3.3 Sensitivity Analysis of the Initial Distributions

Let us now show how the asymptotic behavior is changed when varying the initial conditions. After setting the parameters γ, δ, δ_I, and ε as previously indicated, we select $\alpha = 0.5$ (intermediate progression rate), $\beta = 0.4$, (weak KFc proliferation), $\beta_I = 0.35$ (low proliferation of the ISc), $\lambda = 0.5$ (the probability that a keloid fibroblast cell becomes a malignant cell is not negligible), and we choose the initial distribution $f_i(0, u)$, for $i \in \{1, 2, 3, 4, 5\}$, in the following function set:

$$\mathbf{I_c} = \left\{ g_k(u) = k \exp\left(-100 \left(u - \frac{1}{2} \right)^2 \right) : k = 0, 1, 10, 15 \right\}. \tag{8.12}$$

The functions of $\mathbf{I_c}$ have compact support in the interval $[u_0, u_1] \subseteq [0, 1]$, namely there is a low heterogeneity of the cell activity at the initial instant, and we have depicted their shape in Figure 8.15. Moreover, the following order relation holds: $0 = n[g_0](t) < n[g_1](t) < n[g_{10}](t) < n[g_{15}](t)$.

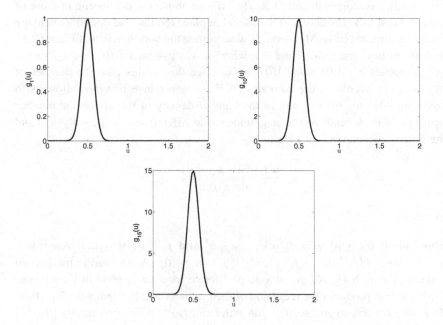

Fig. 8.15   The distributions $g_1(u)$, $g_{10}(u)$, and $g_{15}(u)$.

Each simulation starts when KFc and Mc are not present, namely $f_3(0, u) = f_4(0, u) = g_0(u)$. The analysis refers to the relation among the initial distribution of the NFc, AV, and ISc.

Simulations for $f_1(0,u) = g_{10}(u)$, $f_2(0,u) = g_0(u)$, **and** $f_5(0,u) = g_1(u)$. This assumption means that the activated virus has not reached the wound ($n[f_2](0) = 0$) and after the injury, the number of NFc is greater than the number of the ISc, namely $n[f_5](0) < n[f_1](0)$. We simulate the possible formation of keloid that is triggered only by genetic susceptibility of the patient. As the left panel of Figure 8.16 shows, there is the onset of KFc but not malignant effects, and the dynamics described by the model is just the competition between KFc and ISc. Moreover, KFc show low levels of self-proliferation (see the bottom panel of Figure 8.16).

Simulations for $f_1(0,u) = g_{15}(u)$, $f_2(0,u) = g_{10}(u)$, **and** $f_5(0,u) = g_1(u)$. Thus we are assuming that the number of ISc is less than both the number of NFc and the number of the AV, namely $n[f_5](0) < n[f_2](0) < n[f_1](0)$. As the top panel of Figure 8.17 shows, at the first stage, the sentinel level of the ISc is not able to prevent the free increasing of the number of the AV that, at a certain instant of time $t_{AV} \in [4,5]$, overcomes the number NFc (whose number is decreasing because of the action of AV). The action of the ISc takes place too late and does not avoid the birth of malignant cells. Moreover, If we compare the bottom panel of Figure 8.17 with the bottom panel of Figure 8.1, where it was assumed $f_1(0,u) = f_2(0,u) = g_{10(u)}$, namely $n[f_2](0) = n[f_1](0)$, we can see that in this case the number of Mc does not overcome the number of KFc. Further investigations allow us to conjecture that the overcoming of the number density of the Mc on the number density of KFc depends on the magnitude of the difference $n[f_1](0) - n[f_2](0)$ and the time t_{AV} decreases as the quantity

$$\frac{n[f_1](0) - n[f_2](0)}{n[f_5](0)}$$

increases.

Simulations for $f_1(0,u) = f_2(0,u) = g_{10}(u)$ **and** $f_5(0,u) = g_{15}(u)$. According to this choice $n[f_5](0) > n[f_1](0)$, $n[f_1](0) = n[f_2](0)$. As showed in the bottom panel of Figure 8.18, AV are able to proliferate for a low period of time before the ISc action produces the reduction of their number and the complete depletion. Thus NFc are able to proliferate again at the time $t_{AV} \in [6,7]$ such that $n[f_1](t_{AV}) = n[f_2](t_{AV})$. The sentinel level of the ISc does not allow the self-entities to reach high levels of proliferation or progression (see the distribution functions of the KFc and Mc depicted in Figure 8.19 and compare these distributions with those showed in the bottom panel of Figures 8.3 and 8.4 obtained assuming $f_1(0,u) = f_2(0,u) = g_{10}(u)$ and $f_5(0,u) = g_1(u)$, namely $n[f_5](0) < n[f_1](0)$, $n[f_1](0) = n[f_2](0)$).

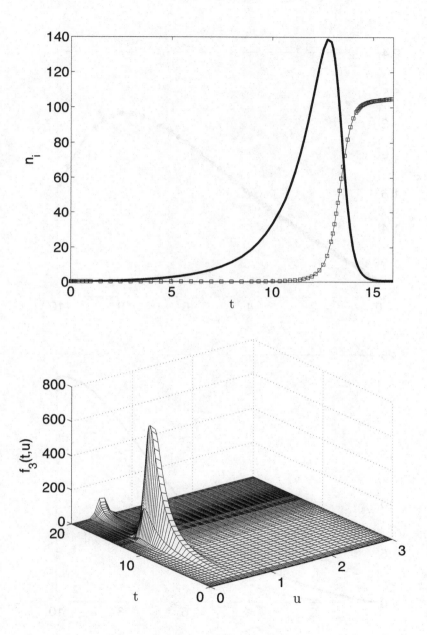

Fig. 8.16 The time evolution of the number density of the KFc (solid line) and Isc (square line) in the top panel and the distribution function of KFc in the bottom panel when $f_1(0,u) = g_{10}(u)$, $f_2(0,u) = g_0(u)$, and $f_5(0,u) = g_1(u)$.

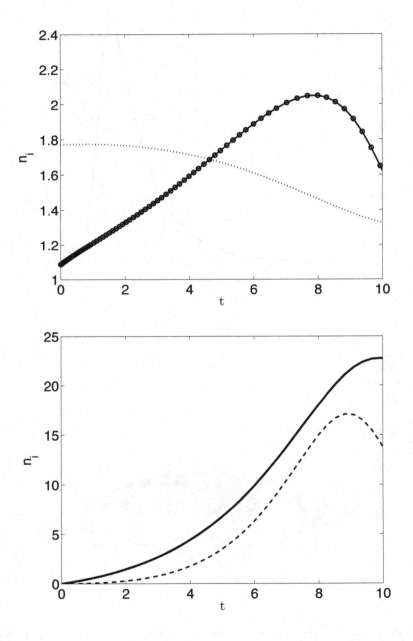

Fig. 8.17   The time evolution, when $f_1(0, u) = g_{15}(u)$, $f_2(0, u) = g_{10}(u)$, and $f_5(0, u) = g_1(u)$, of the density of NFc (dotted line) and AV (circle line) in the top panel, and of the density of the KFc (solid line) and Mc (dashed line) in the bottom panel.

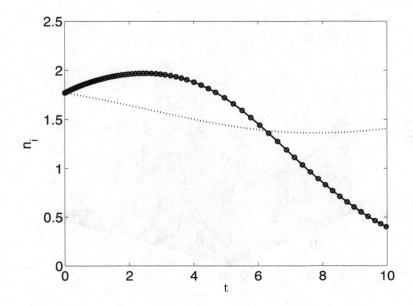

Fig. 8.18   The time evolution of the density of NFc (dotted line) and of the AV (square line) for $f_1(0,u) = f_2(0,u) = g_{10}(u)$ and $f_5(0,u) = g_{15}(u)$.

Simulations for $f_1(0,u) = g_{10}(u)$, $f_2(0,u) = g_{15}(u)$, **and** $f_5(0,u) = g_1(u)$, namely $n[f_5](0) < n[f_1](0) \le n[f_2](0)$. As the top panel of Figure 8.20 shows, where it is stretching out the time simulation to better analyze the behavior, NFc are not able to proliferate because AV proliferate with high velocity until a critical time t_c, where the number of AV equals the number of ISc. After the time t_c, ISc are able to contrast AV. Further investigations allow us to conjecture that the number t_c depends on the quantity

$$\frac{n[f_2](0) - n[f_1](0)}{n[f_1](0) - n[f_5](0)}.$$

The initial action of the AV produces a higher life span of the KFc and Mc (compare the bottom panel of Figure 8.20 with the bottom panel of Figure 8.1 obtained assuming $f_1(0,u) = f_2(0,u) = g_{10}(u)$ and $f_5(0,u) = g_1(u)$, namely $n[f_5](0) < n[f_1](0)$, $n[f_1](0) = n[f_2](0)$) and consequently it starts the inhibition of the ISc (see the top panel of Figure 8.20).

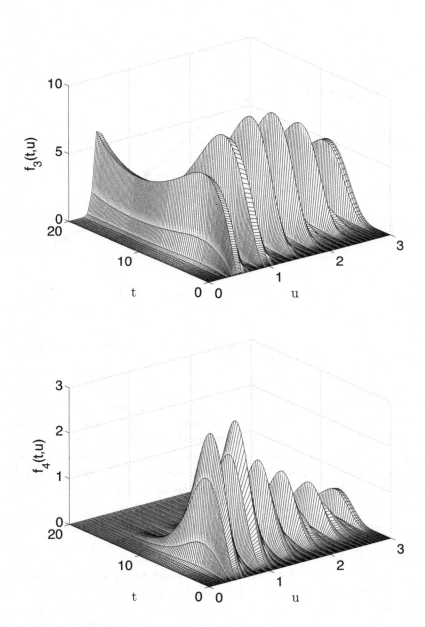

Fig. 8.19   The distribution function of the KFc (top panel) and Mc (bottom panel) for $f_1(0,u) = f_2(0,u) = g_{10}(u)$ and $f_5(0,u) = g_{15}(u)$.

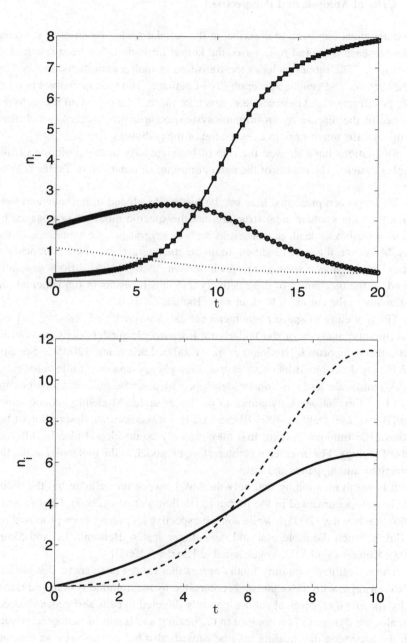

Fig. 8.20 The time evolution of the densities of NFc (dotted line), AV (circle line), and ISc (square line) in the top panel, and of the KFc (solid line) and Mc (dashed line) in the bottom panel, for $f_1(0, u) = g_{10}(u)$, $f_2(0, u) = g_{15}(u)$, and $f_5(0, u) = g_1(u)$.

8.4   Critical Analysis and Perspectives

A mathematical approach to describe, at the cellular scale, the onset of a wound healing disease triggered by a virus, the keloid formation, has been reported in this chapter. The model includes the transition to malignant effects, induced by viruses actions and genetic susceptibility of patients. The system evolves in time, while proliferative and destructive events take place. The organism reacts to the presence of the disease by an immune system competition process, which may end up with the repair or with a regression of the pathology.

Simulations have showed the role of heterogeneity in the evolution of this complex system. The results of the simulations are summarized in Tables 8.2 and 8.3.

The approach presented here can be further developed to improve and generalize both the mathematical structure and the specific model. For instance by including birth and death processes, as well as several genetic mutations of the virus. Moreover, it can be modified, to model the interactions of the therapeutical actions, regarded as a subsystem. As is known, therapeutical actions generally depend on the mechanics of drug delivery and thus it results in fundamental understanding of the nature of the transport [Bianca (2010)].

The modeling of space phenomena can be described, and possibly derived, by asymptotic analysis of the underlying microscopic models delivered by the kinetic theory approach [Bellomo *et al.* (2007); Lachowicz (2005)]. Specific models may describe diffusion, or propagation phenomena with finite velocity.

A multiscale analysis can be developed to describe genetic mutations induced by an evolutionary dynamics at the lower scale. Modeling genetic mutations [Bianca and Delitala (2010)] can lead to a more accurate description of the action of the immune system, that may possibly be developed also in different fields of biology. For instance it is interesting to model, at the molecular scale, the interactions among genes and virus.

It is worth mentioning that only the link between the cellular and the tissue scale is well documented in the literature [Bellomo *et al.* (2007); Chalub *et al.* (2006); Lachowicz (2005)], while some perspective ideas have been proposed for the link between the molecular and the cellular scales [Bellomo, Li, and Maini (2008); Komarova (2007); Vogelstein and Kinzler (2004)].

The scientific community finally agrees that cancer is a genetic disease and that the struggle with the organism is contrasted by the immune system and possibly by specific therapeutical actions possibly directed to cells and genes. Indeed, an analogous dynamics is witnessed in the healing and repair of biological tissues with the exception that healing may be caused, also but not only, by an external action. However, the subsequent dynamics involves all conceivable scales. Indeed, the dynamics moves from the low to the higher scale. Namely the dynamics at the molecular scale, that may be influenced by the interactions with the outer environment, determines the dynamics at the cellular scale. The structure, hence the mathematical models, of tissues depends on the underlying description at the cellular scale.

Table 8.2 The results of the simulations on the sensitivity analysis of the progression rate α of the KFc (first block corresponding to $\alpha = 0.2, 0.5, 0.8$), and of the proliferation rate β_I of the ISc (second block corresponding to $\beta_I = 0, 0.8$).

FIXED PARAMETERS	FREE PARAMETER	SIMULATION RESULTS AND BIOLOGICAL INTERPRETATION
$\beta = 0.4$ $\beta_I = 0.35$ $\gamma = 0.4$ $\lambda = 0.5$ $\delta_I = 0.5$ $\delta = 0.3$ $\varepsilon = 0.5$	$\alpha = 0.2$	Onset of KFc and Mc Mc never overcome KFc Low number of the high activity non-self cells Depletion of the Mc and AV by ISc NFc proliferates again
	$\alpha = 0.5$	Mc overcome NFc High activity levels for KFc and Mc Low number of the AV with high activity NFc are not able to proliferate Partial depletion of the low activity non-self cells
	$\alpha = 0.8$	Mc overcome KFc more and more High proliferation of AV High number of the high activity non-self cells Inhibition of the ISc after competition Destruction of the NFc
$\beta = 0.4$ $\alpha = 0.8$ $\gamma = 0.4$ $\lambda = 0.5$ $\delta_I = 0.5$ $\delta = 0.3$ $\varepsilon = 0.5$	$\beta_I = 0$	Untrammeled increasing of the non-self cells High levels of proliferation of the AV and KFc High levels of progression of the Mc Total inhibition of the ISc Destruction of NFc
	$\beta_I = 0.8$	Depletion of the non-self entities by ISc Revival of the Proliferation of the NFc Rebirth of KFc because genetic susceptibility

Table 8.3   The results of the simulations on the sensitivity analysis of the initial number density $n_i(0)$, for $i \in \{1,2,5\}$, assuming $n_4(0) = n_5(0) = 0$.

INITIAL DISTRIBUTIONS	SIMULATION RESULTS
$f_1(0,u) = g_{10}$, $f_2(0,u) = g_0$, $f_5(0,u) = g_1$	Onset of KFc but not malignant effects Low heterogeneity of the KFc Competition with the ISc and rebirth of KFc
$f_1(0,u) = g_{15}$, $f_2(0,u) = g_{10}$, $f_5(0,u) = g_1$	Initial Free increasing of AV until t_{AV} Onset of KFc and Mc Slow proliferation of the ISc
$f_1(0,u) = f_2(0,u) = g_{10}$, $f_5(0,u) = g_{15}$	Low proliferation of AV Low number of the high activity non-self cells Revival of the proliferation of the NFc
$f_1(0,u) = g_{10}$, $f_2(0,u) = g_{15}$, $f_5(0,u) = g_1$	NFc are not able to proliferate High life span of the KFc and Mc High proliferation of the ISc Depletion of the AV by the ISc

Chapter 9

Macroscopic Models of Chemotaxis by KTAP Asymptotic Methods

9.1 Introduction

This chapter shows how the mathematical tools presented in Chapter 7 can be developed in appropriate manner to derive an important class of macroscopic equations related to chemotaxis phenomena, namely suitable generalizations of the celebrated Keller and Segel model introduced in Chapter 7. As it is well known, chemotaxis consists of the characteristic, movement or orientation of a population (bacteria, cells or other single or multicellular organisms) along a chemical concentration gradient either towards or away from the chemical stimulus (signals), e.g. bacteria swimming to find food, the movement of sperm towards an egg during fertilization, migration of lymphocytes, and cancer metastasis. An extensive literature review on the biological basis of chemotaxis in different contexts has been documented in the review paper [Herrero, Köhn, and Pérez-Pomares (2009)] as well as in the nice *user's guide* proposed in [Hillen and Painter (2009)]. As already mentioned applied mathematicians have been attracted by mathematical problems related to this model as documented by the analysis of the qualitative properties of solutions focusing on different aspects such has hyperbolic behaviors [Herrmann, Niethammer, and Velàzquez (2009); Nadin, Perthame, and Ryzhik (2008); Perthame and Dailbard (2009); Wang and Hillen (2008)], and general qualitative properties of the solutions related also to modifications of the original model [Burger, Dolak-Struss, and Schmeiser (2008); Hillen, Painter, and Scmeiser (2007)], and many others.

Macroscopic models are often derived through a heuristic phenomenological approach, which does not always retain the dynamics at the lower scale where complex events such as competition, growth, and destructive phenomena, occur. The purely phenomenological derivation leaves several questions open. Among others, whether it is reasonable to assume linear diffusion when all the phenomena occur with a finite speed. Moreover, how can nonlinear diffusion phenomena be treated by introducing appropriate models to limit the flux? Therefore, asymptotic methods to derive macroscopic equations from the underlying microscopic description have been developed and were reported in Chapter 7. The contents of this lecture are focused on some specific applications and are developed by referring to the paper [Bellomo *et al.* (2010)], where the aforesaid topic is treated exhaustively.

A prototype to describe the movement of biological populations is the usual linear diffusive (Fokker-Planck) system. However, recent experiments [Callejo *et al.* (2010); Dessaud *et al.* (2007)] have proved that, on the contrary, the propagation of this populations is made by fronts and singularities that are certainly far from that of linear diffusion models. Different ideas have been proposed in order to more clearly understand population propagation, based mainly on nonlinear diffusion or on hyperbolic models that allow front propagations, periodic solutions, breathing modes, singularities, and so on to be transported and preserved. These ideas consist in modifying the model by introducing a nonlinear limited flux of the type

$$\text{div}_{\mathbf{x}} \left(n \frac{\nabla_{\mathbf{x}} n}{\sqrt{n^2 + \frac{v^2}{c^2} |\nabla_{\mathbf{x}} n|^2}} \right)$$

instead of linear diffusion, where $n = n(t, \mathbf{x})$ is the cell density at position $\mathbf{x}$ and time t, v the kinematic viscosity and c the maximum speed of propagation, see [Brenier (2003); Rosenau (1990, 1992)]. The introduction of this type of term can also be motivated by the assumption that particles do not move (diffuse) arbitrarily in the space but, on the contrary, through some privileged curves such as the border of cells. Moreover, the non-physical diffusion is eliminated and the population moves with a finite speed of propagation, c, which is one of the intrinsic characteristics. As a consequence of propagation phenomena in biological tissues, the system behaves more like a hyperbolic system than the usual linear diffusive (Fokker-Planck) system. The dynamical structures, such as propagation fronts, biological responses or stable patterns, are preserved during the time evolution.

Focusing on this topic, let us consider the following macroscopic model for the density and the chemoattractant. This model collects two of the innovating improved terms, with respect to the classical Keller–Segel model, and consists of the choice of a limited flux and of the optimal transport of the cell density $n = n(t, \mathbf{x})$ at position $\mathbf{x}$ and time t according to the chemical signal density $S = S(t, \mathbf{x})$

of the chemo-attractant:

$$
\begin{cases}
\partial_t n = \mathrm{div}_{\mathbf{x}} \left(D_n \dfrac{n \nabla_{\mathbf{x}} n}{\sqrt{n^2 + \frac{D_n^2}{c^2} |\nabla_{\mathbf{x}} n|^2}} - n\chi \dfrac{\nabla_{\mathbf{x}} S}{\sqrt{1 + |\nabla_{\mathbf{x}} S|^2}} \right) + H_2(n, S), \\[4mm]
\partial_t S = \mathrm{div}_{\mathbf{x}}(D_S \cdot \nabla_{\mathbf{x}} S) + H_1(n, S),
\end{cases}
\tag{9.1}
$$

where $H_i(n, S)$, $i = 1, 2$ describes the interactions between both populations, the positive definite terms D_S and D_n are the diffusivity of the chemo-attractant and of the cells, respectively, while $\chi \geq 0$ is the chemotactic sensitivity. It should be pointed out that both modifications are motivated by optimal transportation criteria [Brenier (2003)] that are essential from a qualitative point of view, for instance, for the propagation of singular fronts.

After the above general introduction some details on the contents of this lecture can now be given. Section 9.2 shows how the methodological approach dealt with in Chapter 7, which is summarized in Eqs. (7.16)-(7.18), can be properly specialized to derive models where interactions drive a trend towards equilibrium. These calculations support the contents of Section 9.3 which is devoted to derive, for a mixture of two functional subsystems, various models of chemotaxis. Finally, Section 9.4 proposes a critical analysis, where additional potential developments are examined and presented as research perspectives.

9.2 Linear Turning Kernels: Relaxation Models

This section deals with the derivation of macroscopic equations corresponding to specific characterization of the turning operator $L[f]$. Explicit expression of the macroscopic transport equations, in particular of the source terms, are obtained in the two subsections in the sequel corresponding, respectively, to a scalar system and to a binary mixture. As we shall see, an appropriate choice of such kernel leads to the derivation of interesting models that have been previously proposed on the basis of heuristic arguments, while some of them can be improved, on the basis of the optimal transport theory, by suppressing the description of not physical behaviors such as the blow up of solutions. Of course, the interested reader can propose improved expressions of the kernel with the aim of a deeper investigation of macroscopic behaviors of multicellular systems.

9.2.1 *The Case of a Single Subsystem*

Following the asymptotic methods presented in Subsection 7.3.1, we assume that the velocity space $D_{\mathbf{v}}$ is the $(n-1)$-sphere of radius $r > 0$, namely

$$D_{\mathbf{v}} = r\mathbb{S}^{n-1} = \{\mathbf{v} \in \mathbb{R}^n : \|\mathbf{v}\| = r\}.$$

Moreover, we choose the following kernel:

$$T(\mathbf{v}, \mathbf{v}^*) = \lambda + \beta\, \mathbf{v} \cdot \mathbf{v}^*.$$

The operator $L[f]$ given by Eq. (7.5) can be written, by straightforward calculations, as follows:

$$L[f] = \int_{D_{\mathbf{v}}} \Big((\lambda + \beta\, \mathbf{v} \cdot \mathbf{v}^*) f(\mathbf{v}^*) - (\lambda + \beta\, \mathbf{v} \cdot \mathbf{v}^*) f(\mathbf{v}) \Big) d\mathbf{v}^*$$

$$= \lambda |D_{\mathbf{v}}| \left[\frac{\rho}{|D_{\mathbf{v}}|} \left(1 + \frac{\beta}{\lambda} \mathbf{v} \cdot U \right) - f(\mathbf{v}) \right]. \tag{9.2}$$

It is possible to obtain, by using these calculations, an explicit formula for the function $M_{\rho, \mathbf{U}}$ defined in Assumption 7.2, of Chapter 7, as stated in the following lemma:

Lemma 9.1. *Let $D_{\mathbf{v}} = r\mathbb{S}^{n-1}$ and $L[f]$ be given by formula (9.2). Assume the relation $\beta\, r^2 = \lambda\, n$. Then $L[f]$ verifies Assumptions 7.1 and 7.2 for a function $M_{\rho, U}(\mathbf{v})$ given by*

$$M_{\rho, U}(\mathbf{v}) = \frac{\rho}{|D_{\mathbf{v}}|} \left(1 + \frac{\beta}{\lambda} \mathbf{v} \cdot U \right) = \frac{\rho}{|D_{\mathbf{v}}|} \left(1 + \frac{n}{r^2} \mathbf{v} \cdot U \right), \tag{9.3}$$

and $L[f]$ is the relaxation operator

$$L[f] = \lambda |D_{\mathbf{v}}| \Big(M_{\rho, U}(\mathbf{v}) - f(\mathbf{v}) \Big). \tag{9.4}$$

Moreover, the pressure tensor P_0 defined by Eq. (7.14) associated with $M_{\rho, U}(\mathbf{v})$ is given by:

$$P_0 = \frac{r^2}{n} \rho\, \mathbb{I} - \rho\, U \otimes U. \tag{9.5}$$

Let us now define, for any vector function F defined in D_u, the following scalar quantities:

$$P(F, F) = F(u) \cdot \int_{D_u} F(u^*)\, du^*, \tag{9.6}$$

and

$$C(F, F) = \int_{D_u} \int_{D_u} \mathscr{B}(u_* \to u | u_*, u^*) F(u^*) \cdot F(u_*)\, du_* du^* + F(u) \cdot \int_{D_u} F(u^*)\, du^*.$$

Moreover, for any scalar function f and any vectorial function F both defined in D_u, the following vector quantities:

$$P(f,F) = -\frac{1}{2}\left(f(u) \int_{D_u} F(u^*)\,du^* + F(u) \int_{D_u} f(u^*)\,du^* \right), \qquad (9.7)$$

while the following quantity

$$C(f,F) = \frac{1}{2}\int_{D_u}\int_{D_u} \mathscr{B}(u_* \to u | u_*, u^*)\Big(f(u^*)F(u_*) + f(u_*)F(u^*) \Big)\,du_* du^* + I(f,F),$$

can be defined to generalize in a natural way the definition of operators C and P, so that we preserve the same name for them.

Then, the main result of [Bellomo *et al.* (2010)] can be summarized by the following theorem:

Theorem 9.1. *Let f_ε be a solution of (7.10), with L that of the relaxation model given by (9.2), verifying*

$$\sup_{t \geq 0} \int_\Omega \int_{D_v} \int_{D_u} (f_\varepsilon(t,\mathbf{x},\mathbf{v},u))^p \, du\, d\mathbf{v}\, d\mathbf{x} \leq C < \infty \qquad (9.8)$$

for some $p > 2$, and such that f_ε converges a.e. in $[0,T] \times \Omega \times D_u \times r\mathbb{S}^{n-1}$ for some $T > 0$. We also assume that the kernel $\mathscr{B}$ of the operator C is in $L^2(D_u \times D_u \times D_u)$. Then, the pointwise limit of f_ε is the function $M_{\rho,U}$ given by (9.3), where

$$\rho \equiv \lim_{\varepsilon \to 0} \rho_\varepsilon, \qquad U \equiv \lim_{\varepsilon \to 0} U_\varepsilon.$$

Moreover, in the three presented regimes, the limiting density ρ and velocity U satisfy, respectively:

(1) *If* $\boxed{\delta \geq 0 \text{ and } q > 1}$: *(ρ, U) satisfies the following hyperbolic system without source term:*

$$\begin{cases} \partial_t \rho + div(\rho U) = 0, \\[2mm] \partial_t(\rho U) + \dfrac{r^2}{n}\nabla_x \rho = 0. \end{cases}$$

(2) *If* $\boxed{\delta > 0 \text{ and } q = 1}$: *(ρ, U) satisfies the following hyperbolic system with source term related to conservative interactions:*

$$\begin{cases} \partial_t \rho + div(\rho U) = \dfrac{1}{|D_v|}\left(C(\rho,\rho) + \dfrac{n}{r^2}C(\rho U, \rho U) \right), \\[3mm] \partial_t(\rho U) + \dfrac{r^2}{n}\nabla_x \rho = \dfrac{2}{|D_v|}C(\rho, \rho U). \end{cases}$$

(3) If $\boxed{\delta = 0 \text{ and } q = 1}$: (ρ, U) *verifies the following hyperbolic system whose source term preserves both conservative and proliferating interactions:*

$$\begin{cases} \partial_t \rho + \operatorname{div}(\rho U) = \dfrac{1}{|D_{\mathbf{v}}|}\left(H(\rho,\rho) + \dfrac{n}{r^2}H(\rho U, \rho U)\right), \\[2ex] \partial_t(\rho U) + \dfrac{r^2}{n}\nabla_x \rho = \dfrac{2}{|D_{\mathbf{v}}|}H(\rho, \rho U), \end{cases}$$

where the operator H *is given by* $H := C + P$.

Remark 9.1. The hypothesis $\beta\, r^2 = \lambda\, n$ for the turning (relaxation) operator (9.2) implies essentially the solvability condition

$$\int_{D_{\mathbf{v}}} \mathbf{v}\, L[f]\, d\mathbf{v} = 0.$$

If it is not assumed, the stated regimes (up to some constants) are preserved, but with the addition of a damping term of the form

$$\left(\frac{\beta r^{n+1}}{n} - \lambda r^{n-1}\right)|\mathbb{S}^{n-1}|\,\rho U$$

on the right-hand side of the second equation for the evolution of ρU.

9.2.2   *The Case of a Binary Mixture of Subsystems*

According to the asymptotic methods presented in Subsection 7.3.2 and in the case of the parabolic-parabolic hydrodynamic limit, we consider the following task for the probability kernels:

$$T_1(\mathbf{v}, \mathbf{v}^*) = \sigma_1 M_1(\mathbf{v}), \qquad T_2^0(\mathbf{v}, \mathbf{v}^*) = \sigma_2 M_2(\mathbf{v}), \qquad \sigma_1,\ \sigma_2 > 0.$$

Consequently, the leading turning operators L_1 and L_2^0 become relaxation operators:

$$L_1[g] = -\sigma_1\left(g - \langle g\rangle_{\mathbf{v}} M_1\right), \qquad L_2^0[g] = -\sigma_2\left(g - \langle g\rangle_{\mathbf{v}} M_2\right).$$

In particular, θ_1 and θ_2 are given by

$$\theta_1(\mathbf{v}) = -\frac{1}{\sigma_1}\mathbf{v}M_1(\mathbf{v}), \qquad \theta_2(\mathbf{v}) = -\frac{1}{\sigma_2}\mathbf{v}M_2(\mathbf{v}).$$

According to Theorem 7.1, the following macroscopic equations are derived:

$$\partial_t S - \delta_{p,1}\operatorname{div}_{\mathbf{x}}(D_S \cdot \nabla_{\mathbf{x}}S) = \delta_{q,1}\,\mathscr{C}_1(n,S) + \delta_{q,1}\delta_{r_1,0}\,\mathscr{P}_1(n,S),$$

$$\partial_t n + \operatorname{div}_{\mathbf{x}}(n\,\alpha(S) - D_n \cdot \nabla_{\mathbf{x}}n) = \delta_{q,1}\,\mathscr{C}_2(n,S) + \delta_{q,1}\delta_{r_2,0}\,\mathscr{P}_2(n,S),$$

where $\delta_{a,b}$ stands for the Kronecker delta, the diffusion tensors D_n and D_S are given by

$$D_S = \frac{1}{\sigma_1} \int_{D_v} \mathbf{v} \otimes \mathbf{v} M_1(\mathbf{v}) d\mathbf{v}, \quad D_n = \frac{1}{\sigma_2} \int_{D_v} \mathbf{v} \otimes \mathbf{v} M_2(\mathbf{v}) d\mathbf{v},$$

and

$$\alpha(S) = \frac{1}{\sigma_2} \int_{D_v} \mathbf{v} L_2^1 [M_1 S](M_2)(\mathbf{v}) d\mathbf{v}. \tag{9.9}$$

Moreover, if rotational invariance of the equilibrium distribution, namely $M_i = M_i(|\mathbf{v}|)$ is assumed, the isotropic tensors D_n and D_S are given by:

$$D_S = \left(\frac{1}{3\sigma_1} \int_{D_v} |\mathbf{v}|^2 M_1(\mathbf{v}) d\mathbf{v} \right) I, \quad D_n = \left(\frac{1}{3\sigma_2} \int_{D_v} |\mathbf{v}|^2 M_2(\mathbf{v}) d\mathbf{v} \right) I.$$

9.3 Cellular-Tissue Scale Models of Chemotaxis

The mathematical approach to study chemotaxis was boosted in [Keller and Segel (1971); Keller (1979)]. They introduced a model to study the aggregation of Dictyostelium discoideum due to an attractive chemical substance. The model consists in an advection-diffusion system of two coupled parabolic equations:

$$\begin{cases} \partial_t n = \text{div}_{\mathbf{x}}(D_n \nabla_{\mathbf{x}} n - \chi n \nabla_{\mathbf{x}} S) + H(n,S), \\ \\ \partial_t S = D_S \Delta S + K(n,S), \end{cases} \tag{9.10}$$

where $n = n(t,\mathbf{x})$ is the cell density at position $\mathbf{x}$ and time t, and $S = S(t,\mathbf{x})$ is the density of the chemo-attractant. The positive definite terms D_S and D_n are the diffusivity of the chemo-attractant and of the cells, respectively, while $\chi \geq 0$ is the chemotactic sensitivity. In a more general framework in which diffusions are not isotropic, D_S and D_n could be positive definite matrices.

Applied mathematicians have been attracted by this model due to a variety of challenging problems from the rigorous derivation of the model [Filbet, Laurençot, and Perthame (2005); Chalub *et al.* (2004, 2006); Dolak and Schmeiser (2005)] to the analysis of the blow-up of solutions, among several ones [Andreu *et al.* (2006); Dolbeault and Schmeiser (2009)]. The literature of applied mathematicians to investigate the qualitative properties of this model is documented in [Dolbeault and Schmeiser (2009)] and therein cited bibliography. However, the model needs to be revised to avoid the unrealistic blow up.

It is not completely clear how the term $\text{div}_{\mathbf{x}}(\chi n \nabla_{\mathbf{x}} S)$ induces *per se* the *optimal* movement of the cells towards the pathway determined by the chemoattractant. This term could be modified in a fashion that the flux density of particles is optimized along the trajectory induced by the chemoattractant, namely by minimizing the functional

$$\int \chi n \, dS = \int \chi n \sqrt{1 + |\nabla_{\mathbf{x}} S|^2} \, dx$$

with respect to S, where dS is the measure of the curve defined by S. This approach provides an alternative term in the corresponding Euler-Lagrange equation of type

$$\text{div}_{\mathbf{x}} \left(\chi n \frac{\nabla_{\mathbf{x}} S}{\sqrt{1 + |\nabla_{\mathbf{x}} S|^2}} \right). \qquad (9.11)$$

Of course this term coincides with $\text{div}_{\mathbf{x}}(\chi n \nabla_{\mathbf{x}} S)$ when $|\nabla_{\mathbf{x}} S|$ is very small. However, if $|\nabla_{\mathbf{x}} S| \sim 0$, comparing this scale with the remaining scales of the problem is necessary.

As we mentioned in the introduction, it does not seem realistic to think that cells or bacteria move simply by (linear Fokker-Planck) diffusion, $\text{div}_{\mathbf{x}}(D_n \nabla_{\mathbf{x}} n)$. Other possibilities to modify this approach based on incorporating real phenomena related with cell or bacteria motion (cilium activation or elasticity properties of the membrane, among others) can be considered. For instance, considering a nonlinear limited flux that allows a richer and more realistic dynamics: finite speed of propagation c, preservation of fronts in the evolution, or formation of biological patterns. This is represented by terms of the type

$$\text{div}_{\mathbf{x}} \left(D_n \frac{n \nabla_{\mathbf{x}} n}{\sqrt{n^2 + \frac{D_n^2}{c^2} |\nabla_{\mathbf{x}} n|^2}} \right).$$

Therefore, the model below collects two of the innovating improved terms with respect to the classical model consisting in the choice of a flux limited and in the optimal transport of the population following the chemical signal

$$\begin{cases} \partial_t n = \text{div}_{\mathbf{x}} \left(D_n \dfrac{n \nabla_{\mathbf{x}} n}{\sqrt{n^2 + \frac{D_n^2}{c^2} |\nabla_{\mathbf{x}} n|^2}} - n \chi \dfrac{\nabla_{\mathbf{x}} S}{\sqrt{1 + |\nabla_{\mathbf{x}} S|^2}} \right) + H(n, S), \\[4mm] \partial_t S = \nabla_{\mathbf{x}} (D_S \cdot \nabla_{\mathbf{x}} S) + K(n, S), \end{cases} \qquad (9.12)$$

where n and S denote, respectively, the density of the cells and of the chemoattractant [Bellomo *et al.* (2010)].

A challenging problem consists in the derivation of the model from the underlying description at the cellular scale and, possibly, a revision of the model itself to avoid unrealistic blow up description of phenomena that, in real conditions, show a regular behavior. The derivation needs the two scaling approach reported in Chapter 7. Some examples are given in the sequel.

9.3.1 *Classical Keller-Segel Type Models*

The relaxation kernels presented in Section 9.2.2 together with the choice

$$T_2^1[f_1] = K_{\frac{f_1}{M_1}}(\mathbf{v}, \mathbf{v}^*) \cdot \nabla_{\mathbf{x}} \frac{f_1}{M_1},$$

where $K_{\frac{f_1}{M_1}}(\mathbf{v}, \mathbf{v}^*)$ is a vector valued function, leads to the model

$$L_2^1[M_1 S](M_2) = h(v, S) \cdot \nabla_{\mathbf{x}} S,$$

where

$$h(v, S) = \int_{D_{\mathbf{v}}} \Big(K_S(\mathbf{v}, \mathbf{v}^*) M_2(\mathbf{v}^*) - K_S(\mathbf{v}^*, \mathbf{v}) M_2(\mathbf{v}) \Big) d\mathbf{v}^*.$$

Finally, the quantity $\alpha(S)$ in (9.9) is given by

$$\alpha(S) = \chi(S) \cdot \nabla_{\mathbf{x}} S,$$

where the chemotactic sensitivity $\chi(S)$ is given by the matrix

$$\chi(S) = \frac{1}{\sigma_2} \int_{D_{\mathbf{v}}} \mathbf{v} \otimes h(\mathbf{v}, S) d\mathbf{v}. \tag{9.13}$$

Therefore, the drift term $\text{div}_{\mathbf{x}}(n\,\alpha(S))$ that appears in the macroscopic case stated by Theorem 7.1 becomes:

$$\text{div}_{\mathbf{x}}(n\,\alpha(S)) = \text{div}_{\mathbf{x}}(n\chi(S) \cdot \nabla_{\mathbf{x}} S),$$

which gives a Keller-Segel type model (9.10) in case $p = 1$ of Theorem 7.1.

9.3.2 *Optimal Drift Following the Chemoattractant*

If we combine the relaxation kernels presented in Section 9.2.2 with the following choice for T_2^1:

$$T_2^1[f_1] = K_{\frac{f_1}{M_1}}(\mathbf{v}, \mathbf{v}^*) \cdot \frac{\nabla_{\mathbf{x}} \frac{f_1}{M_1}}{\sqrt{1 + |\nabla_{\mathbf{x}} \frac{f_1}{M_1}|^2}}, \tag{9.14}$$

then, the drift term $\text{div}_{\mathbf{x}}(n\,\alpha(S))$ that appears in the macroscopic cases defined in Theorem 7.1 becomes

$$\text{div}_{\mathbf{x}}(n\,\alpha(S)) = \text{div}_{\mathbf{x}}\left(n\chi(S) \cdot \frac{\nabla_{\mathbf{x}} S}{\sqrt{1 + |\nabla_{\mathbf{x}} S|^2}} \right),$$

where the chemotactic sensitivity $\chi(S)$ is given by the matrix (9.13), and, in general, is not constant. This corresponds to the optimal drift term (9.11) here presented as a modification of the Keller–Segel model (9.10).

The model deduced in Section 9.3.1 can be a reasonable simplification of this one when $|\nabla_{\mathbf{x}} S| \sim 0$, but it is not, in general, a good simplification since the trajectories can develop, for example, spiral patterns.

9.3.3   *Nonlinear Flux-Limited Model by the Mixed Scalings*

The construction of turning operators, that lead to deduce nonlinear flux-limited terms, requires a different approach. Some nonlinear turning operators, obtained from first principles the flux-limited system, here deduced for the hyperbolic-parabolic limit, according to an appropriated choice of the operator L_2. This choice depends of both populations on the drift–diffusion type models analyzed in this section.

Let us briefly discuss how to modify the linear diffusion in order to incorporate optimal criteria for the population transport. To fix the ideas let us consider the very naive example of the heat equation for the evolution of a density of individuals in a population,

$$\partial_t n = \nu \, \Delta n. \tag{9.15}$$

We can rewrite (9.15) as follows:

$$\partial_t n = \mathrm{div}_{\mathbf{x}} \left(n \nabla_{\mathbf{x}} \ln n \right) = \mathrm{div}_{\mathbf{x}} \left(n v \right), \tag{9.16}$$

where $v = \nabla_{\mathbf{x}} \ln n$ is a microscopic velocity associated with individuals.

The heat equation, written as in (9.16), takes the form of a transport kinetic equation, in which the usual parabolic scale $(ht, h^2 \mathbf{x})$ can be viewed as an implicit double (through the velocity) hyperbolic scale $(ht, h\mathbf{x})$. The velocity v is determined, again in a naive way, by both the Fisher entropy of the system, $F(n) = n \ln n$, and the density n,

$$\mathbf{v} = \nabla_{\mathbf{x}} \left(\frac{F(n)}{n} \right). \tag{9.17}$$

To modify the form of the flux in (9.16), we consider a new microscopic velocity, which is the above local velocity (9.17) averaged with respect to the line element associated with the motion of the particle. The velocity (9.17) (in the hyperbolic scale) is taken as the new unit to measure displacements, so that the new velocity is $\nabla_{\mathbf{v}} \sqrt{1 + |\mathbf{v}|^2}$. In this way the velocity can be considered as a measure of the relative entropy in terms of the particle concentration. We thus arrive at a flux limited equation,

$$\partial_t n = \nu \, \mathrm{div}_{\mathbf{x}} \left(\frac{n \nabla_{\mathbf{x}} n}{\sqrt{n^2 + \frac{\nu^2}{c^2} |\nabla_{\mathbf{x}} n|^2}} \right), \tag{9.18}$$

where ν and c are parameters to be fixed; in particular, c represents the maximum macroscopic speed of propagation allowed.

The model was first deduced by [Rosenau (1992)] from different points of view and then derived in [Brenier (2003)] by means of a Monge-Kantorovich mass transport theory as a gradient flow of the Boltzmann entropy

$$\int_{\mathbb{R}^3} \left(\ln(n) - 1 \right) n \, d\mathbf{x}$$

for the metrics corresponding to the cost function

$$k(z) = \begin{cases} c^2 \left(1 - \sqrt{1 - \frac{|z|^2}{c^2}} \right), & \text{if } |z| \le c, \\ +\infty, & \text{if } |z| > c. \end{cases}$$

In order to incorporate this kind of terms in the framework of our kinetic approach (7.27), we proceed by an iterative argument which requires to assume that the kinetic system admits a solution.

Let $k \in \mathbb{N}$ and n_k^ε a given function. Then we define a sequence of operators

$$L_{2,k+1}[f_1^\varepsilon](f_2^\varepsilon) = \int_{D_{\mathbf{v}}} K \left(\mathbf{v}, \mathbf{v}^*, n_k^\varepsilon, \left\langle \frac{f_1^\varepsilon}{M_1} \right\rangle_{\mathbf{v}} \right) f_2^\varepsilon(\mathbf{v}^*) \, d\mathbf{v}^*, \qquad (9.19)$$

where

$$K \left(\mathbf{v}, \mathbf{v}^*, n_k^\varepsilon, \left\langle \frac{f_1^\varepsilon}{M_1} \right\rangle_{\mathbf{v}} \right) = \alpha \left(n_k^\varepsilon, \left\langle \frac{1}{|D_{\mathbf{v}}|} \frac{f_1^\varepsilon}{M_1} \right\rangle_{\mathbf{v}} \right) \mathbf{v} h(\mathbf{v}) - \mathbf{v}^* \mathbf{v} h(\mathbf{v})$$

and

$$\alpha \left(n_k^\varepsilon, \frac{1}{|D_{\mathbf{v}}|} \left\langle \frac{f_1^\varepsilon}{M_1} \right\rangle_{\mathbf{v}} \right) = \nu \frac{\nabla_{\mathbf{x}} n_k^\varepsilon}{\sqrt{\left(n_k^\varepsilon \right)^2 + \frac{\nu^2}{c^2} |\nabla_{\mathbf{x}} n_k^\varepsilon|^2}}$$

$$- \chi \left(\left\langle \frac{1}{|D_{\mathbf{v}}|} \frac{f_1^\varepsilon}{M_1} \right\rangle_{\mathbf{v}} \right) \frac{\nabla_{\mathbf{x}} \left\langle \frac{1}{|D_{\mathbf{v}}|} \frac{f_1^\varepsilon}{M_1} \right\rangle_{\mathbf{v}}}{\sqrt{1 + \left| \nabla_{\mathbf{x}} \left\langle \frac{1}{|D_{\mathbf{v}}|} \frac{f_1^\varepsilon}{M_1} \right\rangle_{\mathbf{v}} \right|^2}}.$$

The way by which the operator $L_{2,k}$ is constructed implies that the hypothesis of Assumption 7.8 hold. Denote by $f_{2,k+1}^\varepsilon$ the solution of the kinetic system (7.47)-(7.48) associated to the above operator (9.19) and $n_{k+1}^\varepsilon = \langle f_{2,k+1}^\varepsilon \rangle_{\mathbf{v}}$. By an appropriate choice of the rest of the operators involved in the linearized kinetic system (7.47)-(7.48), the existence of solutions can be guaranteed for every k by taking the initial condition as $n_{k=0}^\varepsilon$ in order to initialize the sequence. The convergence of the sequence $\{f_{2,k}^\varepsilon\}_k$ to a function f_2^ε, at least weakly in measure, can then be established. In this procedure for the sake of simplicity we have omitted the reference to the k-index for the other population f_1^ε.

Denote by $L_2[f_1^\varepsilon](f_2^\varepsilon)$ the limit as $k \to \infty$ of the set $\{L_{2,k}\}_k$ which satisfies Assumption 7.9. Thus $L_2[f_1^\varepsilon](f_2^\varepsilon)$ is finally defined by

$$K\left(\mathbf{v},\mathbf{v}^*,n^\varepsilon,\left\langle \frac{f_1^\varepsilon}{M_1}\right\rangle_{\mathbf{v}}\right) = \alpha\left(n^\varepsilon,\left\langle \frac{1}{|D_{\mathbf{v}}|}\frac{f_1^\varepsilon}{M_1}\right\rangle_{\mathbf{v}}\right)\mathbf{v}h(\mathbf{v}) - \mathbf{v}^*\mathbf{v}h(\mathbf{v}).$$

Moreover, the limiting *non-linear* current is given by

$$j = \langle \mathbf{v}f_2^0\rangle = n\,\alpha\,(n,S),$$

with

$$\alpha\,(n,S) = v\,\frac{\nabla_{\mathbf{x}}n}{\sqrt{n^2 + \frac{v^2}{c^2}|\nabla_{\mathbf{x}}n|^2}} - \chi\,(S)\,\frac{\nabla_{\mathbf{x}}S}{\sqrt{1+|\nabla_{\mathbf{x}}S|^2}}.$$

Then, we can formally deduce that the limiting system, verified by the macroscopic limiting quantities, is obtained with $\alpha(S)$ replaced by $\alpha(n,S)$:

$$\begin{cases} \partial_t S = \delta_{p,1}\nabla_{\mathbf{x}}(D_S\cdot\nabla_{\mathbf{x}}S) + H_1[n,S], \\[2ex] \partial_t n = \mathrm{div}_{\mathbf{x}}\left(v\,\dfrac{n\nabla_{\mathbf{x}}n}{\sqrt{n^2+\frac{v^2}{c^2}|\nabla_{\mathbf{x}}n|^2}} - n\chi\,\dfrac{\nabla_{\mathbf{x}}S}{\sqrt{1+|\nabla_{\mathbf{x}}S|^2}}\right) + H_1[n,S], \end{cases} \quad (9.20)$$

with

$$H_1[n,S] = \delta_{q,1}\left\langle C_1[\mathbf{f}^0,\mathbf{f}^0]\right\rangle_{\mathbf{v}} + \delta_{q,1}\delta_{r_1,0}\left\langle P_1[\mathbf{f}^0,\mathbf{f}^0]\right\rangle_{\mathbf{v}},$$
$$H_2[n,S] = \delta_{q,1}\left\langle C_2[\mathbf{f}^0,\mathbf{f}^0]\right\rangle_{\mathbf{v}} + \delta_{q,1}\delta_{r_2,0}\left\langle P_2[\mathbf{f}^0,\mathbf{f}^0]\right\rangle_{\mathbf{v}},$$

where $f^0 = (M_1S,f_2^0)$. Then, (9.20) corresponds to the limited flux Keller–Segel model (9.12) with optimal transport of the population n with respect to the chemical signal S.

9.4   Critical Analysis

The applications developed in this chapter have been proposed on the basis of the idea that the phenomenological derivation of tissue macroscopic models based on reaction-diffusion and conservation equations closed by phenomenological material models should be replaced, in the case of biological systems by asymptotic methods from the low to the large scale: genes $\to$ cells $\to$ tissue, as represented in Figure 9.1.

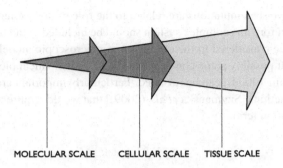

MOLECULAR SCALE   CELLULAR SCALE   TISSUE SCALE

Fig. 9.1   From genes to tissues.

The mathematical tools presented in Chapter 7 have been applied to two specific cases, namely functional subsystems that relax to equilibrium and the derivation of the celebrated Keller and Segel model. The derivation and these two specific applications, that are of interest to the contents of this monograph, have been studied in [Bellomo *et al.* (2010)]. Hopefully the approach can be generalized to a variety of specific cases related to wound healing processes. The key issue consists in modeling both the dynamics at the cellular level and the kernel that characterize the perturbation from the spatially homogeneous case. The general background consists in linking the derivation of tissue models from cellular properties and the underlying description at the cellular scale [Bellomo and Delitala (2008); Xue and Othmer (2009)].

It is worth mentioning, in view of further developments, that the approach defines several analytic problems that constitute interesting research perspectives. The following ones are selected according to the authors' bias:

i) The proof of theorems needs some a *priori* assumptions on the quality of the solutions that can be proved for some specific models;

ii) More general expression of the operator modeling the stochastic jump perturbation can be considered, for instance nonlinearly depending on the distribution function;

iii) Including mutations and therapeutical actions in the line of the general approach of Chapter 7;

iv) Linking the scaling parameters to gene expressions;

v) Treating nonlocal interactions.

As we have seen mutations are related to the role of the immune system, see Chapter 3, therefore this complex system should be included in the functional subsystems that are considered in the derivation of macroscopic models. Finally, let us mention that possibly some classical models of mathematical biology [Murray (2002); Chaplain (2008)] can be revisited. Particularly important are models with nonlocal interactions [Szymanska *et al.* (2009)] that would require a modification of the perturbation term.

Chapter 10

Looking Ahead

10.1 Introduction

This monograph has been devoted to the development of mathematical tools, followed by their application, focusing on the modeling and simulation of complex biological processes under therapeutical and immune defence action. The first part presented the description of the biophysical phenomena under consideration. The second part on the derivation of mathematical tools, and finally the third part deals with applications.

An interesting and challenging research perspective consists in the development of a biological mathematical theory of complex systems. This last chapter develops a critical analysis on this specific topic, which is not, for sure, yet available in the literature due to the lack of first principles. However, applied mathematicians and biologists, develop more and more interdisciplinary projects that look at this ambitious objective, which is considered one of the great challenges of this century

The critical analysis also considers some open problems, which are brought to the attention of both applied mathematicians and biologists, generated through interdisciplinary approaches. Some reasonings on these problems can be useful for the development of mathematical sciences. In fact, some of them exhibit conceptual difficulties that create an attractive research environment. New mathematical tools could be needed to deal with them. Conversely, mathematics could suggest to biologists new research perspectives, possibly by contributing to the experimental investigations in order to identify emerging collective behaviors. Ultimately, the effort to develop a mathematical theory of biological systems could lead to new suggestions on biological theories.

10.2   Some Challenges for Applied Mathematicians and Biologists

The various chapters of this monograph have been devoted to show how a mathematical approach to model complex biological systems by taking advantage of developments the methods of statistical mechanics and kinetic theory. The constantly pursued objective consists in capturing the complexity of biological processes by the qualitative description delivered by models and specifically by their ability to depict emerging behaviors.

It is fascinating objective considering that the scientific community looks intensely at a rigorous formalization of biological sciences by methods of mathematical sciences, similarly to what happened in the past centuries to the interaction between mathematical and physical sciences. Indeed, it is a great challenge for applied mathematician who look for a self-consistent mathematical theory of complex systems and, in particular, biological systems.

Waiting for such a theory, various models have been generated, some of them successful, although in some cases they failed. However, applied mathematicians are getting aware that the application of models to real biological phenomena generates a variety of difficult mathematical problems. An example, among several ones, is offered in the field of cancer modeling reviewed in [Bellomo, Li, and Maini (2008)] and variable number of equations. Specifically, the modeling of the immune competition generates initial value problems for nonlinear integro-differential equations with quadratic type nonlinearities. The main objective of the qualitative analysis consists in showing the large time behavior of the solutions based on the bifurcation parameters that have a role on the two asymptotic trends: blow up of tumor cells related to inhibition of the immune system, and, alternatively activation of the immune system with progressive inhibition of immune cells.

Paper [Bellouquid and Delitala (2005)] and the book [Bellouquid and Delitala (2006)] report about qualitative and computational analysis of such a mathematical problem. The main difficulty is due to the lack of conservation equations, not even mass conservation. Therefore, the result is achieved by sharp inequalities generated by comparing the dynamics at the cellular scale to that at the macroscopic scale giving the time evolution of the number of cells.

The system approach to model complex phenomena in biology offers to applied mathematicians a variety of challenging problems, which need inventing new methods to treat their qualitative and computational analysis.

Interesting challenges are generated by the asymptotic analysis to derive macroscopic models from the underlying description at the microscopic scale. Indeed, all models proposed in this monograph generate problems, which demand further analysis and refinements.

Conversely, biologists can also take advantage of their interactions with mathematical sciences. In fact, still pursuing their interest for experimental research, they can focus the experimental analysis on the tuning of models by identification of the parameters of the various models we have presented in the preceding chapters. Their identification can be achieved by looking at the dynamics of the interaction at the molecular and cellular scale and between these two scales.

Particularly interesting is the multiscale aspect of biological phenomena. For instance the passage from the molecular to the cellular scale needs the assessment of the clustering of genes that produce certain phenomena at the cellular scale. Subsequently, the analysis can be focused on the measurements of gene expression that generate, related to different intensity, emerging phenomena, expression of biological functions, at the cellular scale. Some interesting results are already offered by various recent papers [Gabetta and Ragazzini (2010); Tiuryn, Wójtowicz, and Rudnicky (2007)].

The selection of conceivable experiments should be based on a thoughtful strategy. The next section provides a preliminary analysis of what is still needed to derive a mathematical theory for the class of biological systems under consideration. Experiments can be related to new biological theories that can contribute to such objective.

10.3 How Far is the Mathematical Theory for Biological Systems

This monograph has presented a variety of mathematical tools and a general methodological approach to model complex phenomena in biology. The approach is based on the idea that mathematics should retain the complexity features of biology and that a new system approach has to be developed suitable to reduce complexity, but, at the same time, to retain the essence of the complex dynamics that characterize all biological systems. Hopefully, pursuing the above objectives can lead to mathematical theories in the field of biology.

The mathematical approach proposed in this monograph is based on a new system biology idea, where the overall system is decomposed into functional subsystems (constituted by a large number of interacting entities) characterized by the biological functions collectively expressed by their components. The dynamics of each subsystem is delivered by the kinetic theory of active particles presented in Chapter 6. Modeling interactions among subsystems complete the description of the whole system by mathematical equations.

The mathematical method is required to capture the complexity characteristics described in Chapter 2. The guidelines of the application of the mathematical approach can be summarized referring to them and specifically focusing on the cellular scale:

10.1 *Decomposition* of the overall system into *functional subsystems*, where cells collectively express a common biological function, called *activity*;

10.2 *Representation* of the overall state of each functional subsystem by a probability distribution function over the microscopic state of cells, called active particles;

10.3 *Modeling interactions* among particles within each subsystem and among them to derive an equation suitable to describe the time *evolution* of the above defined distribution function;

10.4 *Modeling interactions at the lower molecular scale* to identify the dynamics that determines the activities at the cellular scale;

10.5 *Derivation of macroscopic models* from the underlying dynamics at the cellular scale by an appropriate development of asymptotic theories, where the intercellular distances are going to zero.

We are aware that this modeling approach leaves several problems still open and it is limited to the cellular dynamics with links to the molecular and tissue scale. For instance it does not treat the analysis at the higher scale, where clustering functional subsystems generate organs and the interactions among organs themselves and with the outer environment. However, focusing on the specific contents of the previous chapters, some preliminary reasonings can be offered concerning the challenging objective of deriving a mathematical theory of complex biological systems.

It can be stated, referring to the above guidelines, that the first three items concern the mathematical approach, which can even be considered a mathematical theory considering that new tools and methods have been invented. Specifically, the decomposition 10.1 is proposed to reduce the complexity of the overall system, whose representation is delivered by stochastic variables according to 10.2. The derivation of the evolution equation for these variables can be obtained by application of the mathematical methods presented in Chapter 6. It is a mathematical theory developed to capture the ten characteristics of complex systems. Some additional reasonings follow.

It is worth stressing that the decomposition into functional subsystems attempts to reduce the complexity of living systems, which are generally constituted by a large variety of components. Indeed, these systems are different from the physical systems analyzed by statistical mechanics, which typically deals with systems containing many copies of a few interacting components. Therefore any mathematical system would not be tractable without an appropriate reduction of this structural complexity.

Moreover, the representation of the state of the system by a distribution function takes into account the presence of a large number of entities, that have been called active particles, constituting each subsystem. The use of a probability distribution models the heterogeneous distribution of their microscopic state. The modeling of ability of active particles to develop specific strategies, namely of a self-organization strategy, depending on the state of the surrounding environment is due to the selection of the activity variable.

Interactions are modeled by stochastic games where the output is related to their strategy often depending on surviving and adaptation ability finalized to their own individual or collective fitness. Interactions involve immediate neighbors, but in some cases, also distant ones considering that biological systems have the ability to communicate and may, in some cases, choose different observation paths. Nonlinear interactions can be modeled according to [Bellomo, Bianca, and Mongiovi (2010)].

Darwinian type mutations caused by successive selections, that in some cases can be rapid, of entities, which become progressively resistent to a mutating environment are modeled by transition into new functional subsystems characterized by different phenotype properties.

Time is a key variable in the model due to different rates characterizing biological processes. Accordingly small changes at the scale of genes can lead to large effects at the cellular scale, and consequently at that of tissue. This dynamics is also related to feedbacks from their environments.

A multiscale approach has been developed by linking the dynamics of a cell to the molecular (genetic) level that determines the cellular behaviors, while models of tissues have been related to the underlying description at the cellular scale. Ultimately, the structures of the macroscopic tissues depend on gene expression.

However, the mathematical theory needs to be further developed. Some recent results can be mentioned. For instance a detailed analysis of the role of space dynamics is treated in [Bianca (2010)], while fragmentation and aggregations (clustering) phenomena are dealt with in [Bellouquid and Bianca (2010)]. The main issue consists in the modeling of nonlinear interactions to take into account the topological distribution in the space of microscopic state of the field particles that interact with the test and candidate particles.

Therefore, the mathematical theory needs further developments to improve its predictive ability. In particular, the main objective remains the characterization of the interaction terms η_{hk}, B_{hk}, and μ_{hk}^{i} involving active particles, for instance cells or genes. The key issue consists in treating interactions whose dynamics is related to that at the lower scale, namely to the higher or lower expression of genes. Constructing this link means moving from models to a proper theory.

The analogy with the inert matter is immediate although substantial differences can be put in evidence. A specific analogy can be given with the physics of classical particles whose dynamics is ruled by particle interaction models described by attractive-repulsive potentials. Newtonian mechanics provides the necessary mathematical background to describe particle interactions by attraction-repulsion potentials of the interacting particles, or by mechanical collisions which preserve mass, momentum and energy. It is worth stressing that a deep analysis of the inner structure of atoms or molecules is not necessary, but simply a theoretical description of the interaction potentials which govern pair interactions between particles will suffice. In the case of multicellular systems, the cell state includes, in addition to the mechanical state, biological functions which have the ability of modifying their mechanical behavior. In our case, biology should contribute, by experiments and theoretical interpretations, to describe the outcome of cellular interactions. Specifically, the above four terms should be elucidated at the molecular level, i.e., at the lower scale.

Although, at present, such a theory is not yet available, it is well understood that the objective can be achieved only through a detailed analysis of gene expression related to biological functions at the molecular scale. This research topic has been dealt with by, among others, [Nowak and Sigmund (2004); Komarova and Wodarz (2004); Komarova (2006, 2007); Gatenby, Vincent, and Gillies (2005)], related to specific theories in the field of biological sciences, which give evidence of the evolutionary aspect, referred to gene expression, of several, may be almost all, biological phenomena.

The various theoretical approaches known in the literature postulate probabilistic models of gene expression, while gene interactions among themselves and with the external environment should be taken into account. A robust theory represents a challenging and highly attractive objective.

The passage from a theory valid in the field of mathematical sciences to a theory valid also in the field of biology can be achieved by a deep insight into biological sciences to obtain the afore mentioned terms by a theoretical rather than phenomenological approach. Experiments, possibly new ones still to be designed, and theory can march together to pursue this objective. At present this objective is a challenging target that will ask for a great deal of research activity in the next decades.

10.4 Closure

The reasonings of the preceding section have been focused on the state of the art and perspectives concerning the assessment of the theoretical approach, both biological and mathematical, to the class of systems under consideration. Some implications can be extracted by them on the way of organizing the research activity in the field.

For instance, let us consider, as a conceivable closure to this monograph, the main aspect of the interactions between mathematical and biological sciences. Namely, we consider the conceptual paths that link Biology to Mathematics. Some thoughts can be developed starting from the following *dilemma*:

Dilemma: Should mathematics attempt to reproduce experiments by equations whose parameters are identified on the basis of empirical data? Or, in alternative, should mathematics develop new structures and methods, hopefully a new theory, suitable to capture the complexity of the biological phenomena and finally base experiments on theoretical foundations?

This dilemma is always object of discussions and even contrasts that oppose different opinions within the community of applied mathematicians. Of course the answer is related to the personal bias and experience of the researchers involved in the dilemma.

Personal opinion: The conflict is not wise considering that both conceptual approaches should march together.

It is plain that a mathematical theory of biological systems needs new tools and possibly a new mathematics. Indeed, this is consistent with the continuous development of mathematical sciences whenever it has been motivated by physical and applied sciences. This means a different way to look at experiments, which can be used to validate a theory, rather than using mathematics to reproduce them.

The scientific community is finally aware that the modeling of complex systems by simple mathematics is too naive to be realistic and effective. When simple models are used, even if they appear to be successful after the identification of parameters, it often happens that these parameters are simply numbers that are not precisely linked to the essence of the biological system under consideration. The assessment of parameters should be related to the molecular and cellular scale, while models are often derived at large scales.

Therefore, we wish to stress that the alternative, which pursues the objective of developing a proper mathematical theory is definitely related to the essence of the multiscale essence of biological phenomena. Accordingly, looking for this alternative leads to a deeper understanding of real phenomena and hence of experiments.

Possibly, mathematical models can depict emerging behaviors that are not even observed in experiments. Subsequently, new experiments can be designed to improve the understanding of biological phenomena. When this event occurs, one can state that the interaction between mathematics and biology has been successful and that it consists in the first step towards a mathematical theory of complex biological systems, that is the great challenge of this century.

Appendix A

Mathematical Modeling of Space and Velocity-Dependent Systems

A.1 Introduction

This appendix is focused on the modeling of systems of active particles where the space and velocity variables have a fundamental role. The derivation of mathematical structures is developed firstly in the case of homogeneous (or constant) distribution over the activity variable, namely when the activity variable is the same for all particles (Section A.2) and finally when interactions modify both mechanical and activity variable (Section A.3). The approach proposed in this appendix is a technical development of that proposed in Chapter 6.

A.2 Mathematical Tools for Homogeneous Activity Systems

This section deals with the derivation of mathematical structures suitable to describe the evolution of large systems of interacting active particles belonging to different functional subsystems. The derivation is developed in the case of homogeneous (or constant) distribution over the activity variable, namely when the activity variable is the same for all particles. Interactions modify the mechanical variables, for instance position and velocity, while their activity is not modified. Still, the activity variable has a remarkable role on the dynamics of the interactions.

Let us first consider the case of one functional subsystem only, the generalization to several interaction subsystems is, as we shall see, immediate. Interactions, at the time t, refer, as usual, to three types of particles:

- *Test* particles with microscopic state, at time t, defined by the variable $(\mathbf{x}, \mathbf{v})$, whose distribution function is $f = f(t, \mathbf{x}, \mathbf{v})$.

- *Field* particles with microscopic state, at time t, defined by the variable $(\mathbf{x}^*, \mathbf{v}^*)$,

179

whose distribution function is $f^* = f(t, \mathbf{x}^*, \mathbf{v}^*)$.

• *Candidate* particles with microscopic state, at time t, defined by the variable $(\mathbf{x}_*, \mathbf{v}_*)$, whose distribution function is $f_* = f(t, \mathbf{x}_*, \mathbf{v}_*)$.

Two different types of interactions are considered: *conservative interactions*, when particles modify their microscopic state, and *nonconservative interactions*, when interactions generate proliferation or destruction of particles in their microscopic state.

The mathematical framework refers to the evolution in time and space of the test particle. The derivation for f is based on the following balance of equations in the elementary volume of the phase space:

$$\frac{df}{dt} d\mathbf{x} d\mathbf{v} = \left(G[f] - L[f] + P[f] \right) d\mathbf{x} d\mathbf{v}, \tag{A.1}$$

where interactions of candidate and test particles refer to the field particles, and:

$G[f] = G[f](t, \mathbf{x}, \mathbf{v})$ denotes the *gain* of candidate particles into the state $(\mathbf{x}, \mathbf{v})$;

$L[f] = L[f](t, \mathbf{x}, \mathbf{v})$ models the *loss* of test particles with state $(\mathbf{x}, \mathbf{v})$;

$P[f] = P[f](t, \mathbf{x}, \mathbf{v})$ models *proliferation/destruction* of test particles.

Therefore, a basic aspect of the characterization of the above equation is the modeling of interactions among particles. Bearing all the above in mind, the modeling is developed on the basis of the following assumptions:

H.A.1. The candidate and/or test particles in $\mathbf{x} \in \Omega = \Omega(x)$, with velocity $\mathbf{v}_*$ and $\mathbf{v}$, respectively, interact with the field particles in $\mathbf{x}^*$, with velocity $\mathbf{v}^*$ located in its interaction domain Ω, $\mathbf{x}^* \in \Omega$.

H.A.2. Interactions are weighted by a suitable term $\eta[\rho] = \eta[\rho](t, \mathbf{x}^*)$, that can be interpreted as an *interaction rate*, which depends on the local density ρ in the position $\mathbf{x}^*$ of the field particles.

H.A.3. The distance and topological distribution of the intensity of the interactions is weighted by a function $w = w(\mathbf{x}, \mathbf{x}^*)$ such that:

$$\int_\Omega w(\mathbf{x}, \mathbf{x}^*) d\mathbf{x}^* = 1. \tag{A.2}$$

H.A.4. The candidate particle f_* modifies its velocity $\mathbf{v}_*$ according to the probability density denoted by $\mathscr{A}(\mathbf{v}_* \to \mathbf{v} | \mathbf{v}, \mathbf{v}^*)$, which represents the probability that a candidate particle, with velocity $\mathbf{v}_*$, reaches, in the same position, the velocity $\mathbf{v}$ after an interaction with a field particle with velocity $\mathbf{v}^*$. The density satisfies the following property:

$$\int_{D_\mathbf{v}} \mathscr{A}(\mathbf{v}_* \to \mathbf{v} | \mathbf{v}_*, \mathbf{v}^*) d\mathbf{v} = 1, \qquad \forall \mathbf{v}_*, \mathbf{v}^* \in D_\mathbf{v}, \tag{A.3}$$

while the test particle loses its state $\mathbf{v}$ after interactions with field particles with velocity $\mathbf{v}^*$. In general, $\mathscr{A}$ is not consistent with conservation of momentum and energy. In fact the presence of an activity variable (although the same for all active particles) implies the onset of a behavior, which modifies the classical conservation laws.

H.A.5. The candidate particle, in $\mathbf{x}$, can proliferate, or be destroyed, with net birth/death rate $\mu(\mathbf{x}, \mathbf{x}^*)$, due to encounters with field particles in $\mathbf{x}^*$.

The result derived by Assumptions H.A.1–H.A.5 is as follows:

$$(\partial_t + \mathbf{v} \cdot \partial_\mathbf{x})\, f(t, \mathbf{x}, \mathbf{v}) = (G[f] - L[f] + P[f])(t, \mathbf{x}, \mathbf{v}), \qquad (A.4)$$

where

$$G[f] = \int_\Lambda \eta[\rho] w(\mathbf{x}, \mathbf{x}^*) \mathscr{A}(\mathbf{v}_* \to \mathbf{v} | \mathbf{v}_*, \mathbf{v}^*) f(t, \mathbf{x}, \mathbf{v}_*) f(t, \mathbf{x}^*, \mathbf{v}^*)\, d\mathbf{x}^* d\mathbf{v}_* d\mathbf{v}^*, \qquad (A.5)$$

$$L[f] = f(t, \mathbf{x}, \mathbf{v}) \int_\Gamma \eta[\rho] w(\mathbf{x}, \mathbf{x}^*) f(t, \mathbf{x}^*, \mathbf{v}^*)\, d\mathbf{x}^* d\mathbf{v}^*, \qquad (A.6)$$

and

$$P[f] = f(t, \mathbf{x}, \mathbf{v}) \int_\Gamma \eta[\rho] w(\mathbf{x}, \mathbf{x}^*) \mu(\mathbf{x}, \mathbf{x}^*) f(t, \mathbf{x}^*, \mathbf{v}^*)\, d\mathbf{x}^* d\mathbf{v}^*, \qquad (A.7)$$

where $\Lambda = \Omega \times D_\mathbf{v} \times D_\mathbf{v}$ and $\Gamma = \Omega \times D_\mathbf{v}$.

It is worth pointing out that some models are such that the density $\mathscr{A}$ is conditioned also by the distribution function of the field particle.

The generalization to a system of several interacting functional subsystems is immediate. It is simply necessary to include interactions of active particles for a large system of n functional subsystems labeled by the index $i \in \{1, \ldots, n\}$. Technical calculations yield:

$$(\partial_t + \mathbf{v} \cdot \partial_\mathbf{x})\, f_i(t, \mathbf{x}, \mathbf{v}) = \sum_{j=1}^n \left(G_{ij}[\mathbf{f}] - L_{ij}[\mathbf{f}] + P_{ij}[\mathbf{f}] \right)(t, \mathbf{x}, \mathbf{v}), \qquad (A.8)$$

where

$$G_{ij}[\mathbf{f}] = \int_\Lambda \eta_{ij}[\rho_j] w_{ij}(\mathbf{x}, \mathbf{x}^*) \mathscr{A}_{ij}(\mathbf{v}_* \to \mathbf{v} | \mathbf{v}_*, \mathbf{v}^*) f_i(t, \mathbf{x}, \mathbf{v}_*) f_j(t, \mathbf{x}^*, \mathbf{v}^*)\, d\mathbf{x}^* d\mathbf{v}_* d\mathbf{v}^*, \qquad (A.9)$$

$$L_{ij}[\mathbf{f}] = f_i(t, \mathbf{x}, \mathbf{v}) \int_\Gamma \eta_{ij}[\rho_j] w_{ij}(\mathbf{x}, \mathbf{x}^*) f_j(t, \mathbf{x}^*, \mathbf{v}^*)\, d\mathbf{x}^* d\mathbf{v}^*, \qquad (A.10)$$

and

$$P_{ij}[\mathbf{f}] = f_i(t, \mathbf{x}, \mathbf{v}) \int_\Gamma \eta_{ij}[\rho_j] w_{ij}(\mathbf{x}, \mathbf{x}^*) \mu_{ij}(\mathbf{x}, \mathbf{x}^*) f_j(t, \mathbf{x}^*, \mathbf{v}^*)\, d\mathbf{x}^* d\mathbf{v}^*. \qquad (A.11)$$

A further development refers to modeling the dynamics of particles, which move from a functional subsystem to another one, as sketched in Figure 6.4. In this case the mathematical structure is developed as follows:

$$(\partial_t + \mathbf{v} \cdot \partial_{\mathbf{x}}) \, f_i(t, \mathbf{x}, \mathbf{v}) = \sum_{h=1}^{n} \sum_{k=1}^{n} \left(G_{hk}^i[\mathbf{f}] + P_{hk}^i[\mathbf{f}] \right)(t, \mathbf{x}, \mathbf{v}) - \sum_{j=1}^{n} L_{ij}[\mathbf{f}](t, \mathbf{x}, \mathbf{v}),$$

(A.12)

where

$$G_{hk}^i[\mathbf{f}] = \int_\Lambda \eta_{hk}[\rho_k](t, \mathbf{x}^*) w_{hk}(\mathbf{x}, \mathbf{x}^*) \, \mathscr{A}_{hk}(\mathbf{v}_* \to \mathbf{v}, h \to i \,|\, \mathbf{v}_*, \mathbf{v}^*) \times$$
$$f_h(t, \mathbf{x}, \mathbf{v}_*) f_k(t, \mathbf{x}^*, \mathbf{v}^*) \, d\mathbf{x}^* \, d\mathbf{v}_* \, d\mathbf{v}^*,$$

(A.13)

and

$$P_{hk}^i[\mathbf{f}] = f_h(t, \mathbf{x}, \mathbf{v}) \int_\Gamma \eta_{hk}[\rho_k] w_{hk}(\mathbf{x}, \mathbf{x}^*) \, \mu_{hk}^i(\mathbf{x}, \mathbf{x}^*) f_k(t, \mathbf{x}^*, \mathbf{v}^*) \, d\mathbf{x}^* \, d\mathbf{v}^*,$$

(A.14)

where $\mathscr{A}_{hk}^i(\mathbf{v}_* \to \mathbf{v}, h \to i \,|\, \mathbf{v}_*, \mathbf{v}^*)$ and $\mu_{hk}^i(\mathbf{x}, \mathbf{x}^*)$ denote, respectively, the probability density of transition into the state $(\mathbf{x}, \mathbf{v})$ of the functional subsystem i, and the proliferation/destruction net birth/death rate into the functional subsystem i, due to the encounter with the particles belonging to the functional subsystems h and k.

Analogous reasonings and structures can be developed in the case of discrete space and velocity variables, which can be seen either as a discrete variable or as a method to identify functional subsystems.

The frameworks proposed in this section refer to a system, which does not interact with the outer environment. These actions are often necessary in modeling systems in life sciences.

A.3   Mathematical Tools for Heterogeneous Activity Systems

This section is focused on the modeling of systems of active particles when interactions modify both mechanical and activity variable. Namely, the modeling of stochastic, topological games is developed and, subsequently, the evolution equation for the distribution function over the state of the particles is derived using a balance of particles in the elementary volume of the space of the microscopic states.

In principles, the above program is simply a technical development of the methods used in Chapter 6. However, rather than looking for a very general structure, it is useful, for modeling purposes, looking for suitable specializations. Therefore, let us focus on systems such that:

Conjecture A.1: *Interactions modify the activity variable according to the topological stochastic games introduced in Section A.1, however independently on the distribution of the velocity variable, modification of the velocity of the interacting particles depends also on the activity variable.*

The above conjecture is based on the idea that particles develop an individual strategy that depends on the activity expressed by the other particles, but not by the velocity distribution.

Let us now focus (in the case of one functional subsystem) on the interactions of a candidate or a test particle in $\mathbf{x}$ (with velocities $\mathbf{v}_*$, $\mathbf{v}$, and activity u_*, u, respectively) with the field particles in $\mathbf{x}^*$, with velocity $\mathbf{v}^*$ and activity u^* located in its interaction domain Ω, namely $\mathbf{x}^* \in \Omega$. We assume, as in Section A.1, that interactions are weighted by the term $\eta[\rho] = \eta[\rho](t, \mathbf{x}^*)$ interpreted as an ***interaction rate***, while the distance and topological distribution of the intensity of the interactions is weighted by the function $w = w(\mathbf{x}, \mathbf{x}^*)$. In details, the following assumption characterizes these interactions:

H.A.5. The candidate particle modifies its state according to the probability density $\mathscr{A}$, which has the following factorization:

$$\mathscr{A}(\mathbf{v}_* \to \mathbf{v}, u_* \to u | \mathbf{v}_*, \mathbf{v}^*, u_*, u^*) = \mathscr{B}(u_* \to u | u_*, u^*) \mathscr{C}(\mathbf{v}_* \to \mathbf{v} | \mathbf{v}_*, \mathbf{v}^*, u_*, u^*),$$
(A.15)

where $\mathscr{A}$ denotes the probability density that a candidate particles with state $(\mathbf{v}_*, u_*)$ reaches the state $(\mathbf{v}, u)$ after an interaction with the field particles with state $(\mathbf{v}^*, u^*)$, while the test particle loses its state $\mathbf{v}$ and u after interactions with field particles with velocity $\mathbf{v}^*$ and activity u^*.

The quantities defined in Assumption H.A.5 are probability densities, therefore the following properties hold true:

$$\int_{D_\mathbf{v} \times D_u} \mathscr{A}(\mathbf{v}_* \to \mathbf{v}, u_* \to u | \mathbf{v}_*, \mathbf{v}^*, u_*, u^*) \, d\mathbf{v} \, du = 1, \qquad \forall \mathbf{v}_*, \mathbf{v}^* \in D_\mathbf{v}, \forall u_*, u^* \in D_u,$$
(A.16)

while after the factorization (A.15):

$$\int_{D_u} \mathscr{B}(u_* \to u | u_*, u^*) \, du = 1, \qquad \forall u_*, u^* \in D_u,$$
(A.17)

and

$$\int_{D_\mathbf{v}} \mathscr{C}(\mathbf{v}_* \to \mathbf{v} | \mathbf{v}_*, \mathbf{v}^*, u_*, u^*) \, d\mathbf{v} = 1, \qquad \forall \mathbf{v}_*, \mathbf{v}^* \in D_\mathbf{v}, \qquad \forall u_*, u^* \in D_u.$$
(A.18)

Equating, analogously to the approach of the preceding section, the total variation of f in the elementary volume of the space of the microscopic states yields:

$$(\partial_t + \mathbf{v} \cdot \partial_{\mathbf{x}}) f(t, \mathbf{x}, \mathbf{v}, u) = \big(G[f] - L[f] + P[f]\big)(t, \mathbf{x}, \mathbf{v}, u), \qquad (A.19)$$

where

$$G[f] = \int_{\Lambda \times D_u \times D_u} \eta[\rho](t, \mathbf{x}^*) w(\mathbf{x}, \mathbf{x}^*) \mathscr{B}(u_* \to u \,|\, u_*, u^*) \mathscr{C}(\mathbf{v}_* \to \mathbf{v} \,|\, \mathbf{v}_*, \mathbf{v}^*, u_*, u^*)$$
$$\times f(t, \mathbf{x}, \mathbf{v}_*, u_*) f(t, \mathbf{x}^*, \mathbf{v}^*, u^*) \, d\mathbf{x}^* \, d\mathbf{v}_* \, d\mathbf{v}^* \, du_* \, du^*, \qquad (A.20)$$

$$L[f] = f(t, \mathbf{x}, \mathbf{v}, u) \int_{\Gamma \times D_u} \eta[\rho](t, \mathbf{x}^*) w(\mathbf{x}, \mathbf{x}^*) f(t, \mathbf{x}^*, \mathbf{v}^*, u^*) \, d\mathbf{x}^* \, d\mathbf{v}^* \, du^*, \quad (A.21)$$

and

$$P[f] = f(t, \mathbf{x}, \mathbf{v}, u) \int_{\Gamma \times D_u} \eta[\rho] \, \mu(\mathbf{x}, \mathbf{x}^*, u_*, u^*) \, w(\mathbf{x}, \mathbf{x}^*) f(t, \mathbf{x}^*, \mathbf{v}^*, u^*) \, d\mathbf{v}^* \, du^* \, d\mathbf{x}^*.$$
$$(A.22)$$

Of course, if Conjecture A.1 is not applied, one simply has to suppress the factorization and leave the more general expression of $\mathscr{A}$ instead of the product of $\mathscr{B}$ and $\mathscr{C}$. The conjecture can simply be used for modeling purposes. As already mentioned these densities can, in some specific cases, be conditioned by the distribution function of the field particles.

The generalization to the case of several interacting subsystems, as well as to systems with discrete velocities, is technical. One has to consider interactions with all functional subsystems, in the first case, and with all discrete velocities, in the second case, precisely as we have seen in Section A.1. Detailed calculations are reported only in the general case, which includes transition from one functional subsystem to another. The corresponding mathematical structure, with meaning of notations analogous to that we have seen in Section A.1, is as follows:

$$(\partial_t + \mathbf{v} \cdot \partial_{\mathbf{x}}) f_i(t, \mathbf{x}, \mathbf{v}, u) = \sum_{h=1}^{n} \sum_{k=1}^{n} \Big(G_{hk}^i[\mathbf{f}] + S_{hk}^i[\mathbf{f}] \Big)(t, \mathbf{x}, \mathbf{v}, u) - \sum_{j=1}^{n} L_{ij}[\mathbf{f}](t, \mathbf{x}, \mathbf{v}, u),$$
$$(A.23)$$

where

$$G_{hk}^i[\mathbf{f}] = \int_{\Lambda \times D_u \times D_u} \eta_{hk}[\rho_k] w_{hk}(\mathbf{x}, \mathbf{x}^*) \, \mathscr{B}_{hk}(u_* \to u, h \to i \,|\, u_*, u^*)$$
$$\times \mathscr{C}_{hk}(\mathbf{v}_* \to \mathbf{v} \,|\, \mathbf{v}_*, \mathbf{v}^*, u_*, u^*) f_h(t, \mathbf{x}, \mathbf{v}_*, u_*)$$
$$\times f_k(t, \mathbf{x}^*, \mathbf{v}^*, u^*) \, d\mathbf{x}^* \, d\mathbf{v}_* \, d\mathbf{v}^* \, du_* \, du^*, \qquad (A.24)$$

$$P_{hk}^i[\mathbf{f}] = \int_{\Gamma \times D_u \times D_u} \eta_{hk}[\rho_k](t, \mathbf{x}^*) w_{hk}(\mathbf{x}, \mathbf{x}^*) \, \mu_{hk}^i(\mathbf{x}, \mathbf{x}^*, u_*, u^*) f_h(t, \mathbf{x}, \mathbf{v}, u_*)$$
$$\times f_k(t, \mathbf{x}^*, \mathbf{v}^*, u^*) \, d\mathbf{x}^* \, d\mathbf{v}^* \, du_* \, du^*, \qquad (A.25)$$

and

$$L_{ij}[\mathbf{f}] = f_i(t,\mathbf{x},\mathbf{v},u) \int_{\Gamma \times D_u} \eta_{ij}[\rho_j](t,\mathbf{x}^*) w_{ij}(\mathbf{x},\mathbf{x}^*) f_j(t,\mathbf{x}^*,\mathbf{v}^*,u^*) d\mathbf{x}^* d\mathbf{v}^* du^*.$$

(A.26)

In some specific applications, for instance the modeling of immune competition in multicellular systems, motivate to study the particular cases in which active particles have a velocity distribution that is constant in time and that is not modified by interactions. Moreover, the distribution of particles is uniform in space. In this case, the only interactions that play a role are those involving the activity variable. Therefore, performing the integration with respect to the velocity variable yields:

$$\partial_t f_i(t,u) = \sum_{h=1}^{n} \sum_{k=1}^{n} \int_{D_u \times D_u} \eta_{hk}[\rho_k] \mathscr{B}_{hk}(u_* \to u, h \to i | u_*, u^*)$$
$$\times f_h(t,u_*) f_k(t,u^*) du_* du^*$$
$$+ \sum_{h=1}^{n} \sum_{k=1}^{n} \int_{D_u \times D_u} \eta_{hk}[\rho_k] \mu_{hk}^i(u_*, u^*; u) f_h(t,u_*) f_k(t,u^*) du_* du^*$$
$$- f_i(t,u) \sum_{k=1}^{n} \int_{D_u} \eta_{ik}[\rho_k] f_k(t,u^*) du^*.$$

(A.27)

It is worth observing that some models are such that the dependent variables $\mathbf{x}$ and $\mathbf{v}$ have no real physical meaning in the sense of mechanical sciences. This aspect induces further technical modifications of the mathematical structure. For instance, the modeling of the term η may include dependence on the state of the interacting active particles.

The above modeling approach, used in the case of closed system, can be generalized to open systems where interactions with the outer environment occur. Moreover, as already mentioned in Part III nonlinear interactions need to be taken into account in various case. This difficult topic has been treated in the case space homogeneity [Bellomo, Bianca, and Mongiovi (2010)], but not in the space dependent case.

Glossary

- **Active particles.** Biological entities, for instance cells, whose microscopic state includes, in addition to geometrical and mechanical variables, also an additional variable, called activity, which represents the individual ability to express a specific strategy.
- **Activity.** Biological functions expressed by active particles.
- **Adaptive innate immunity.** Ability of the immune system toward an antigen or agent bearing an antigen that is expressed in the absence of a prior exposure of an organism to this specific agent or cell.
- **Agranulocytes.** Cells without granules. This class includes lymphocytes and monocytes.
- **AIDS**. Acronym of "acquired immunodeficiency syndrome" caused by infection by the human immunodeficiency virus HIV.
- **Allele.** Gene defined by the phenotype that it creates or by the proteins that it produces.
- **Anastomosis.** Natural, or surgically created, junction between two blood vessels.
- **Angiogenesis.** Process by which new blood vessels are formed.
- **Angiogenesis factor.** Growth factor that induces angiogenesis.
- **Angiogenesis inhibitor.** Agent, for instance endostatin and angiostatin, that antagonize or even inhibit the development of new blood vessels.
- **Angiostatin.** Fragment of plasminogen with multiple anti-angiogenic activities.
- **Antibody.** Any protein generally produced by specialized B cells after stimulation by an antigen and acting specifically against the antigen in the immune response. Antibodies are also referred to as immunoglobulins and gamma-globulins.
- **Antigen.** Molecule or portion of a molecule that can be recognized and bound

by an antibody.

- **Apoptosis.** Program of cellular self-destruction, namely programmed death.
- **Autoimmune.** Process in which the immune system attacks an organism's own normal cells or tissues.
- **B-cells.** Cells which rise and mature in the bone marrow, and, after activation, differentiate into plasma cells, which secrete free-floating antibodies.
- **B lymphocytes** For short B cells, a class of lymphocytes, whose antigen receptor is a cell-surface immunoglobulin molecule. If activated by an antigen B cells differentiate into plasma cells producing antibody with the same specificity of their initial receptor.
- **Bacteria.** Single-celled organisms that have no nucleus. They are completely independent organisms able to eat and reproduce very quickly under favorable conditions.
- **Basement membrane.** Thin layer of extracellular matrix between the epithelium and its connective tissue.
- **Bone marrow.** Organ in the hollow center of bones.
- **Bone wound healing.** Fracture healing process that can be essentially divided into three parts: Reactive phase, reparative phase and remodeling phase.
- **Capillary.** The finest branch of blood vessels connecting small arteries and small veins.
- **Capillary sprouts.** New capillary blood vessels formed by growth and migration of endothelial cells from existing blood vessels.
- **Cell clock.** Network of signaling proteins in the nucleus that regulate and orchestrate progression of the cell through the cell cycle.
- **Cell cycle.** Sequence of changes if a cell from when it is generated by cell division to when it ends with formation of a daughter cell.
- **Cell differentiation.** The process by which a cell acquires a specialized function.
- **Cell maturation.** The stage between division and differentiation during which the specialized components grow and make the cell ready to perform functions that are typical of its type.
- **Cellular reaction.** Occurs when the inner lining of the blood vessel (endothelium) is damaged, that results in the exposure of the connective tissue (collagen) around it to the circulating blood (collagen exposure).
- **Chemotactic substance.** Any substance that induces chemotaxis.
- **Chemotaxis.** Directed movement of a microorganism or cell in response to a chemical stimulus.
- **Clone** Identical cell or DNE molecule descended from a single progenitor.
- **Corneal wound healing.** The epithelial layer contains some growth factors

among which we can find the epidermal growth factor, that seems to be the main chemical which regulates the balance between cell loss and proliferation.

- **Cytokine.** Extracellular signaling protein that acts as a local mediator in cell-cell communication.
- **Cytoplasm.** Viscous contents of a cell contained in the plasma membrane outside its nucleus.
- **Dermis.** Inner or deeper layer of the skin.
- **Differentiation.** Process by which a cell acquires a specialized phenotype that typical of a tissue.
- **Dendritic cell** Immune cell that phacocytoses fragments of cells or infectious agents.
- **Dormancy.** A period during which an incipient tumor remains in a steady state.
- **Dysplasia.** Pre-malignant tissue of abnormal cells.
- **Effector.** An agent that carries out a biological process rather than simply regulating it.
- **Endothelium.** Single sheet of flattened cells that forms the lining of all blood vessels.
- **Enzymes.** Proteins that catalyze a specific chemical reaction.
- **Epithelial cells.** Cells of a layer that forms the linking of a cavity or duct. A specialization includes the skin.
- **Epigenetic.** Changes in the behavior of a cell that is not related to DNA modifications.
- **Epidermis.** Outermost layer of the skin that forms a protective barrier against the external environment.
- **Expression.** Transcription of an active gene or synthesis of a protein from it.
- **Expression program.** Coordinated expression of a series of genes.
- **Extracellular matrix.** Secreted proteins that surround cells within the tissue and create structures in the intercellular space.
- **Fibrosis or scar tissue.** Breakdown of normal tissue dominates in fibrosis through the collagen secretion of fibroblasts. Destruction of the basal membrane on top of the destruction of parenchyma cells leads to fibrosis.
- **Functional genomics.** Technology to gauge cell phenotypes by measuring the expression level of multiple genes.
- **Functional subsystem.** Collection of active particles, which have the ability to express the same activity.
- **Gammaglobulin.** See antibody.
- **Gene amplification.** Increase of the number of copies of a gene.
- **Gene family.** Group of genes that descend evolutionary from a common an-

cestral gene.

- **Gene pool.** Collection of genes present in the genomes of all members of a species.
- **Genetic background.** The entire array of alleles carried in a genome with the exception of a small number of genes that are the subject of study.
- **Genotype.** Overall genetic constitution of an organism.
- **Granulocyte.** Type of Leukocyte that contains granules in the cytoplasm. Granulocytes are composed of three cell types: neutrophils, eosinophils, and basophils.
- **Growth factor.** Extracellular signaling molecule that stimulates cell proliferation.
- **Haptotaxis.** Orientation movement in response to a stimulus provided by contact with a solid body.
- **Heterogeneity of active particles.** The characteristics of active particles or agents differs for entities with the same biological structure, for instance due to different phenotype expression generated by the same genotype.
- **Histocompatibility.** Ability of tissue cells, inserted into a host organism, to be tolerated by the immune system of the host.
- **Humoral reaction.** It occurs by activation of thrombocytes results in the release of more mediators from their granulae which further support the vaso-constriction.
- **Hyperplasia.** Abnormal accumulation of normal appearing cells in a tissue.
- **Hypoxia.** Reduction of oxygen levels.
- **Immune response.** Response of the body to an antigen. It results in the formation of antibodies and cells with the ability to react to the antigen to render it harmless.
- **Immune response - cellular.** Ability of specific cell types such as cytotoxic T cells, natural killer cells, macrophages, to recognize and identify abnormal cells and infectious cells.
- **Immune suppression.** Suppression, due to any cause, of the natural ability to the immune response.
- **Immune system.** System of the body that protects it by producing the immune response.
- **Immunodeficiency.** Inability to produce a sufficient number of antibodies and cells with the ability to react to the antigen. It occurs when a part of the immune system is not present or is not working properly.
- **Immunoglobulin.** See antibody.
- **Immuno-therapy.** Treatment to induce the immune response.
- **Inhibitors.** Natural or synthetic substances capable of stopping or evel slow-

ing down biological processes.

- **Integrin.** Protein involved in the adhesion of cells to the extracellular matrix.
- **Interferon.** Group of proteins produced by cells in response to a virus, a parasite in the cell, or a chemical. Interferon can limit the replication of a virus in cells.
- **Interleukin.** Growth and differentiation factor that stimulates cells of the immune system.
- **Invasion.** Process by which cells, for instance tumor cells, move from one site to the adjacent tissues.
- **Keloid.** Increased collagen synthesis will lead to a hypertrophic scar.
- **Leukocyte.** White or colored cell with a nucleus found in tissues and blood.
- **Liver Fibrosis.** Fibro-proliferative disease where the pathological wound healing response results in hepatocytes necrosis and apoptosis.
- **Low-expression.** Expression of an RNA or protein at a level lower than the normal.
- **Lymph.** Interstitial fluid between cells that is drained by lymph nodes to lymphatic vessels that form the lymphatic system.
- **The Lymphatic System.** Network of lymphatic vessels.
- **Macrophage.** Scavenger cell that engulfs cellular debris and damaged cells by phagocytosis.
- **Mathematical (phenomenological) models.** Equations that, implemented with suitable initial and/or boundary conditions, can describe the behavior of a system. Tuning of parameters of the model are obtained by empirical data.
- **Mathematical theory of biological systems.** Equations related to a well defined mathematical structure that, implemented with suitable initial and/or boundary conditions, can describe the behavior of a broad class of systems. Parameters of the model are obtained by a robust biological theory.
- **Metastasis.** Transfer of health-impairing agency, for instance cancer, to a new site of the body.
- **Mitosis.** Process in the nucleus of a dividing cell and results in the formation of two new nuclei each of them has the same number of chromosomes.
- **Modules.** Components of biological systems within the approach of system biology.
- **Monocyte.** White blood cells produced in the bone marrow that migrate to tissues and mature as macrophages.
- **Morphogenesis.** Creation of a shape, for instance in tissues.
- **Motility.** Spontaneous movement, generally referred to individual cells.
- **Necrosis Factor.** Factor that induces local death of body tissues.
- **Neoplasia.** Uncontrolled growth of tissue.

- **Network.** System constituted by several interacting functional subsystems. A network can constitute, as a specific case, an organ.
- **Neutrophil.** Granulocyte that expresses Fc receptors and has the ability of recognizing various types of infectious agents, specifically bacteria.
- **Oncogene.** A gene that induce cancer.
- **Over-expression.** Expression of an RNA or protein at a level higher than the normal.
- **Phenotype.** Observable trait of an organism.
- **Progression.** Applies to tumor cells to indicate the multi-step sequential evolution of normal to cancer cells.
- **Protease.** Enzyme that degrades some proteins, for instance collagen.
- **Protein.** Any of nitrogen-containing substances that consists of chain of amino acids present in all living cells.
- **Receptor.** Protein that is capable of binding a signaling molecule. Receptors can emit signals, for instance inducing proliferation.
- **Regeneration.** Tissue recovery dominates in regeneration through proliferation of the parenchyma cells that are already present.
- **Regulatory element.** Part of a gene that regulates its expression.
- **Repression.** Regulatory mechanism that shut down the expression of a gene.
- **Systems biology.** It is a new field of biological sciences that aims at developing a system-level understanding of biological systems.
- **Stem cell.** Cell that is capable of self-renewal and of generating a daughter cell that develop new phenotypes. In some cases the phenotypes are more differentiated than those of the mother cell.
- **Stochastic event.** Output that occurs with a certain probability.
- **Synapse.** Point at which a nervous impulse passes from one neuron to the other.
- **T cells.** Any of several subset of lymphocyte defined by their development in the thymus and by heterodimeric receptors associate with CD3 proteins, namely a polypeptide complex associated with the receptor and functions in signal transduction.
- **T helper cells.** Cells tha stimulate the growth and differentiation of B-cells.
- **T regulatory cells.** Cells that modulate the immune response.
- **T suppressor cells.** Cells that stop the immune response when the infection has been defeated.
- **Th lymphocyte.** A T cell type that secrets numerous cytokines that play an important role in the immunity.
- **Thymus.** Glandular organ largely composed by lymphoid tissue and localized in the neck region. It contributes to the development of the immune system.

- **Transgene.** A cloned gene inserted into the germ line of a vertebrate.
- **Tumor angiogenic factor.** A substance that generates proliferation of new blood vessels.
- **Tumor associated macrophages.** Macrophages to solid tumor sites.
- **Tumor Necrosis Factor.** A hormone produced by macrophages that is able to kill tumor cells, and it also promotes the creation of new blood vessels.
- **Tumor rejection.** Process by which an organism prevents, generally by the action of the immune system, the formation of a tumor.
- **Vascularization.** Formation of blood vessels, such as capillary sprouts, in a tissue.
- **Vascular reaction.** Occurs when small damaged blood vessels (arterioles) contract reflexively in two directions: by drawing back into the tissue (retraction) and by contraction of the blood vessel itself which closes the vessel (reflexive vasoconstriction).
- **Virus.** Particle fragment of DNA (or RNA) in a protective coat. The virus comes in contact with a cell, attaches itself to the cell wall and injects its DNA (and perhaps a few enzymes) into the cell.
- **White blood cells.** Leukocytes.
- **Wound.** Rupture of the natural cohesion of tissues.
- **Wound healing, or wound repair.** Process by which the skin, or some other organ, repairs itself after injury.

Bibliography

P. Alonso, L. Rioja, and C. Pera, Keloids: A viral hypothesis, *Medical Hypotheses*, **70** (2008) 156-166.

W. Alt and A. Deutsch, **Dynamics of Cell and Tissue Motion**, Birkhauser, Boston, (1997).

S.M. Anderton and D.C. Wraith, Selection and fine-tuning of the autoimmune T-cell repertoire. *Nature Reviews Immunology*, **2** (2002) 487-498.

F. Andreu, V. Caselles, J.M. Mazón, and S. Moll, Finite Propagation Speed for Limited Flux Diffusion Equations, *Archives Rational Mechanics Analysis*, **182** (2006) 269-297.

L. Arlotti, N. Bellomo, and E. De Angelis, Generalized kinetic (Boltzmann) models: Mathematical structures and applications, *Mathematical Models and Methods in Applied Sciences*, **12** (2002) 567-591.

L. Arlotti, A. Gamba, and M. Lachowicz, A kinetic model of tumor/immune system cellular interaction, *J. Theoretical Medicine*, **4** (2002) 39-50.

E.J. Arnsdorf, P. Tummala, and C.R. Jacobs, Non-canonical Wnt signaling and N-Catherin related β-catening signaling play a role in mechanically induced osteogenic call fate, *Plos One*, **4** (2009) e5388.

P. Auger, R. Bravo de la Parra, J.C. Poggiale, E. Sanchez, and L. Sanz, Aggregation methods in dynamical systems in population and community dynamics, *Physics of Life Reviews*, **5** (2008) 79-105.

C.B. Basbaum and Z. Werb, Focalized proteolysis: spatial and temporal regulation of extracellular matrix degradation at the cell surface, *Current Opinion Cell Biology*, **8** (1996) 731-738.

N. Bellomo, **Modelling Complex Living Systems - A Kinetic Theory and Stochastic Game Approach**, Birkhäuser, Boston, (2008).

N. Bellomo, Modeling the hiding-learning dynamics in large living systems *Applied Mathematical Letters*, **23**, (2010), 907-911.

N. Bellomo and A. Bellouquid, From a class of kinetic models to macroscopic equations for multicellular systems in biology, *Discrete Continuous Dynamical System B*, **4** (2004) 59-80.

N. Bellomo and A. Bellouquid, On the onset of nonlinearity for diffusion models of binary mixtures of biological materials by asymptotic analysis, *International J. Nonlinear Mechanics*, **41** (2006) 281-293.

N. Bellomo and A. Bellouquid, On the derivation of macroscopic tissue equations from hybrid models of the kinetic theory of multicellular growing systems - The effect of global equilibrium, *Nonlinear Analysis Hybrid Systems*, **3** (2009) 215-224.

N. Bellomo, A. Bellouquid, and M.A. Herrero, From microscopic to macroscopic description of multicellular systems and biological growing tissues, *Computers Mathematics with Application*, **53** (2007) 647-663.

N. Bellomo, A. Bellouquid, J. Nieto, and J. Soler, Multicellular growing systems: Hyperbolic limits towards macroscopic description, *Mathematical Models and Methods in Applied Sciences*, **17** (2007) 1675-1693.

N. Bellomo, A. Bellouquid, J. Nieto, and J. Soler, Complexity and mathematical tools toward the modelling of multicellular growing systems, *Mathematical Computer Modelling*, **51** (2010) 441-451.

N. Bellomo, A. Bellouquid, J. Nieto, and J. Soler, Multiscale biological tissue models and flux-limited chemotaxis from binary mixtures of multicellular growing systems, *Mathematical Models and Methods in Applied Sciences*, **20** (2010) 1179-1207.

N. Bellomo, C. Bianca, and M. Delitala, Complexity analysis and mathematical tools towards the modelling of living systems, *Physics of Life Reviews*, **6** (2009) 144-175.

N. Bellomo, C. Bianca, and M.S. Mongiovi, On the modeling of nonlinear interactions in large complex systems, *Applied Mathematics Letters*, **6** (2010) 144-175.

N. Bellomo, E. De Angelis, and L. Preziosi, Multiscale modeling and mathematical problems related to tumor evolution and medical therapy, *J. Theoretical Medicine*, **5** (2003) 111-136.

N. Bellomo, and M. Delitala, From the mathematical kinetic, and stochastic game theory to modelling mutations, onset, progression and immune competition of cancer cells, *Physics of Life Reviews*, **5** (2008) 183-206.

N. Bellomo and M. Delitala, On the coupling of higher and lower scales by the mathematical kinetic theory of active particles, *Applied Mathematical Letters*, **22** (2008) 646-650.

N. Bellomo and G. Forni, Complex multicellular systems and immune competition: New paradigms looking for a mathematical theory, *Current Topics in Developmental Biology*, **81** (2008) 485-502.

N. Bellomo, N.K. Li, and P.K. Maini, On the foundations of cancer modelling: selected topics, speculations, and perspectives, *Mathematical Models and Methods in Applied Sciences*, **18** (2008) 593-646.

A. Bellouquid and C. Bianca, Modelling aggregation-fragmentation phenomena from kinetic to macroscopic scales, *Mathematical and Computer Modelling*, **52** (2010) 802-813.

A. Bellouquid and M. Delitala, Mathematical methods and tools of kinetic theory towards modelling complex biological systems, *Mathematical Models and Methods in Applied Sciences*, **15** (2005) 1639-1666.

A. Bellouquid and M. Delitala, **Modelling Complex Biological Systems - A Kinetic Theory Approach**, Birkhäuser, Boston, (2006).

C. Bianca, On the modelling of space dynamics in the kinetic theory for active particles, *Mathematical and Computer Modelling*, **51** (2010) 72-83.

C. Bianca, Mathematical modelling for keloid formation triggered by virus: malignant effects and immune system competition, *Mathematical Models and Methods in Ap-

plied Sciences, **21(2)** (2010) doi: 10.1142/S021820251100509X.

C. Bianca, On the mathematical transport theory in microporous media: the billiard approach, *Nonlinear Analysis Hybrid Systems*, **4** (2010) 699-735.

C. Bianca and M. Delitala, Genetic mutations and immune system competition: A model by the kinetic theory of active particles, *Preprint*, (2010), to be published.

L. Bonilla and J. Soler, High field limit for the Vlasov–Poisson–Fokker–Planck system: a comparison of different perturbation methods, *Mathematical Models and Methods in Applied Sciences*, **11** (2001) 1457-1681.

Y. Brenier, Extended Monge-Kantorovich Theory, in **Optimal Transportation and Applications**, Lectures given at the C.I.M.E. Summer School help in Martina Franca, L.A. Caffarelli and S. Salsa (eds.), Lecture Notes in Math. 1813, Springer-Verlag, (2003) 91-122.

C.T. Brighton and R.M. Hunt, Histochemical localization of calcium in the fracture callus with potassium pyroantimonate: possible role of chondrocyte mitochondrial calcium in callus calcification, *Journal of Bone and Joint Surgery*, **68-A** (1986) 703-715.

M. Burger, Y. Dolak-Struss, and C. Schmeiser *Communications in Mathematical Sciences*, **6** (2008) 1-28.

A. Callejo, E. Mollica, A. Bilioni, N. Gorfinkiel, C. Torroja, L. Doglio, J. Sierra, J. Ibenez, and I. Guerrero I., A cytoneme-mediated baso-lateral transport, controlled by Dispatched, iHog and Dally-like, operates in Hedgehog gradient formation, (2010), preprint.

C. Cattani and A. Ciancio, Hybrid two scales mathematical tools for active particles modeling complex systems with learing hiding dynamics, *Mathematical Models and Methods in Applied Sciences*, **17** (2007) 171-188.

C. Cattani and A. Ciancio, Qualitative analysis of second-order models of tumor-immune system competition, *Mathematical and Computer Modelling*, **47** (2008) 1339-1355.

F.A. Chalub, P. Markowich, B. Perthame, and C. Schmeiser, Kinetic models for chemotaxis and their drift-diffusion limits, *Monatshefe für Mathematik*, **142** (2004) 123-141.

F.A. Chalub, Y. Dolak-Struss, P. Markowich, D. Oeltz, C. Schmeiser, and A. Soref, Model hierarchies for cell aggregation by chemotaxis, *Mathematical Models and Methods in Applied Sciences*, **16** (2006) 1173-1198.

M.A.J. Chaplain, Modeling aspects of cancer growth: Insight from mathematical, and numerical analysis and computational simulations, Lecture Notes in Mathematics 1940, (2008) 147-200.

A. Chauviere, L. Preziosi, and C. Verdier, Eds., **Cell Mechanics**, CRC Press, Boca Raton, (2010).

R.A.F. Clark, Overview and general consideration of wound repair, in *The molecular and Cellular Biology of Wound Repair*, R.A.F. Clark and P.H. Henson, Eds., Plenum, New York, (1989) 3-34.

R.A.F. Clark, Wound repair, *Current Opinion Cell Biology*, **1** (1989) 1000-1008.

R.A.F. Clark, Biology of dermal wound repair, *Dermatology Cli.*, **11** (1993) 647-666.

E.L. Cooper, Evolution of immune system from self/not self to danger to artificial immune system, *Physics of Life Reviews*, **7** (2010) 55-78.

L. Corrias and B. Perthame, Asymptotic decay for the solutions of the parabolic-parabolic Keller-Segel chemotaxis systems in critical spaces, *Mathematical and Computer Modelling*, **47** (2008) 755-764.

E. De Angelis, M. Delitala, A. Marasco, and A. Romano, Bifurcation analysis for a mean field modelling of tumor and immune system competition, *Mathematical Computer Modelling*, **37** (2003) 1131-1142.

E. De Angelis and P.E. Jabin, Qualitative Analysis of a mean field model of tumor-immune system competition, *Mathematical Models and Methods in Applied Sciences*, **13** (2003) 187-206.

E. De Angelis and P.E. Jabin, Mathematical models of therapeutical actions related to tumour and immune system competition, *Mathematical Methods in Applied Sciences*, **28** (2005) 2061-2083.

E. Dessaud, L.L. Yang, K. Hill, B. Cox, F. Ulloa, A. Ribeiro, A. Mynett, B.G. Novitch, and J. Briscoe, Interpretation of the sonic hedgehog morphogen gradient by a temporal adaptation mechanism, *Nature*, **450** (2007) 717-720.

O. Dieckmann and J.A.P. Heesterbeck, **Mathematical Epidemiology of Infectious Diseases**, Wiley, New York, (2000).

Y. Dolak and C. Schmeiser, Kinetic models for chemotaxis: Hydrodynamic limits and spatio temporal mechanisms, *J. Mathematical Biology*, **51** (2005) 595-615.

J. Dolbeault and C. Schmeiser, The two-dimensional Keller-Segel model after blow-up, *Discrete and Continuous Dynamical Systems*, **25** (2009), 109-121.

R. Dover and N.A. Wright, The cell proliferation kinetics of the epidermis: In **Physiology, Biochemistry and Molecular Biology of the Skin** (L.A. Goldsmith Ed.), New York, Oxford University Press, (1991) 239-265.

K. Drucis, M. Kolev, W. Majda, and B. Zubik-Kowal, Nonlinear modeling with mammographic evidence of carcinoma, *Nonlinear Analysis Real World Applications*, **10** (2010) 4326-4334.

R.S. English and P.D. Shenefelt, Keloids and hypertrophic scars, *Dermatol Surgery*, **25** (1999) 631-638.

R. Erban, and H.G. Othmer, From individual to collective behaviour in chemotaxis, *SIAM J. Applied Mathematics*, **65** (2004) 361-391.

F. Filbet, P. Laurençot, and B. Perthame, Derivation of hyperbolic models for chemosensitive movement, *J. Mathematical Biology*, **50** (2005) 189-207.

J. Forrester, A. Dick, P. McMenamin, and W. Lee, **The Eye**, W.B. Saunders Ltd, London, (1996).

S.A. Frank, **Dynamics of Cancer: Inheritance, and Evolution**, Princeton University Press, (2007).

M. Frank, M. Herty, and M. Schäfer, Optimal treatment planning in radiotherapy based on Boltzmann equation, *Mathematical Models and Methods in Applied Sciences*, **18** (2008) 573-952.

L. Fusi, Macroscopi models for fibroproliferative disorders: A review, *Mathematical and Computer Modelling*, **50** (2009) 1474-1494.

E. Gabetta and E. Ragazzini, About the gene families size diustribution ina recent model of genome evolution, *Mathematical Models and Methods in Applied Sciences*, **20** (2010) 1005-1020.

R.A. Gatenby, T.L. Vincent, and R.J. Gillies R.J., Evolutionary dynamics in carcinogenesis, *Mathematical Models and Methods in Applied Sciences*, **15** (2005) 1619-1638.

M.B. Gerstein, C. Bruce, J.S. Rozowsky, D. Zheng, J. Du, J.O. Korbel, O. Emanuelsson, Z.D. Zhang, S. Weissman, and M. Snyder, What is a gene, post-ENCODE? History and updated definition, *Genome Research*, **17** (2007) 669-681.

I. Gipson, Adhesive mechanisms of the corneal epithelium, *Acta Ophthalmica*, **70** (Suppl.) (1992) 13-17.

I. Gipson and T. Inatomi, Extra-cellular matrix and growth factors in corneal wound healing, *Current Opininion Ophthalmica*, **6** (1995) 3-10.

B. Goldstein, J.R. Faeder, and W.S. Hlavaceck, Mathematical and computational models of immune-receptor signalling, *Nature Reviews, Immunology*, **4** (2004) 445-456.

A.W. Ham and W.R. Harris, **Repair and Transplantation of Bone, The Biochemistry and Physiology of Bone**, New York: Academic Press, (1972) 337-399.

D. Hanahan and R.A. Weinberg, The hallmarks of cancer, *Cell*, **100** (2000) 57-70.

H.L. Hartwell, J.J. Hopfield, S. Leibner, and A.W. Murray, From molecular to modular cell biology, *Nature*, **402** (1999) c47-c52.

A. Hastings and M.A. Palmer, A bright future for biologists and mathematicians?, *Science*, **299**, No. 5615, (2003) 2003-2004.

A. Hastings, P. Arzberger, B. Bolker, S. Collins, A.R. Ives, N.A. Johnson, and M.A. Palmer, Quantitative bioscience for the 21st century, *Bioscience*, **55** (2005) 511-517.

E. Hay *et al.*, N-Catherin interacts with axin and LRP5 to negatively regulate Wnt/β-catenin signaling, osteoblast function and bone formation, *Molecular and Cellular Biology*, **29** (2009) 953-964.

M. Herrero, On the role of mathematics in biology, *J. Mathematical Biology*, **54** (2007) 887-889.

M.A. Herrero, A. Köhn, and J.M. Pérez-Pomares, Modelling vascular morphogenesis: Current view on blood vessels development, *Mathematical Models and Methods in Applied Sciences*, **19** (2009) 1483-1538.

M. Herrmann, B. Niethammer, and J.J.L. Velàzquez, Self-similar solutions of the LSW model with encounters, *J. Differential Equations*, **247** (2009) 2282-2309.

T. Hillen and H. Othmer, The diffusion limit of transport equations derived from velocity jump processes, *SIAM J. Applied Mathematics*, **61** (2000) 751-775.

T. Hillen and K.J. Painter, A user's guide to chemotaxis, *J. Mathematical Biology*, **58** (2009) 183-217.

T. Hillen, K.J. Painter, and C. Scmeiser, Global for chemotaxis with finite radius, *Discrete and Continuous Dynamica Systems B*, **7** (2007) 125-144.

J.D. Hunprey and K.R. Rajagopal, A constrained mixture model of growth and remodeling of soft tissues, *Mathematical Models and Methods in Applied Sciences*, **12** (2002) 407-430.

International Human Genome Sequencing Consortium, Finishing the euchromatic sequence of the human genome, *Nature*, **431** (2005) 931-945.

J.P. Iredale, R.C. Benyon, J. Pickering, M. McCullen, M. Northrop, S. Pawley, C. Hovell, and M.J. Arthur, Mechanisms of spontaneous resolution of rat liver fibrosis. Hepatic stellate cell apoptosis and reduced hepatic expression of metalloproteinase inhibitors. *J. Clinical Investigation*, **102** (1998) 538-549.

Z. Jackiewicz, C.L. Jorcyk, M. Kolev, BV. Zubik-Kowal, Correlation between animal and mathematical models for prostate cancer progression, *Computational and Mathematical Methods in Medicine*, **10** (2009) 241-252.

R.W. Jennings and T.K. Hunt, Overview of post-natal wound healing, **Foetal Wound Healing**, (N.S. A Adzick & M.T. Longaker Eds), New York, Elsevier, (1992) 25-52.

K. Kang, A. Stevens, and J.J.L. Velàzquez, Qualitative behavior of a Keller-Segel model with non-diffusive memory, *Communications in Partial Differential Equations*, **35** (2010) 245-274.

E.F. Keller and L.A. Segel, Model for chemotaxis, *J. Theoretical Biology*, **30** (1971) 225-234.

E.F. Keller, Assessing the Keller-Segel model: how has it fared?, **Biological Growth and Spread (Proc. Conf., Heidelberg, 1979)**, Springer Berlin, 1980 379-387.

R.C.P. Kerckhoffs, S.N. Healy, T.P. Usyk, and A.D. McCulloch, Computational methods for cardiac electrophysiology. *Proceedings IEEE*, **94** (2006) 769-783.

R.S. Kirsner and W.H. Eaglstein, The wound healing process, *Dermatology Clinics*, **11** (1993) 629-640.

H. Kitano, Perspectives on systems biology, *New Generation Computing*, **18** (2000) 199-216.

M. Kolev, Mathematical modelling of the competition between tumors and immune system considering the role of antibodies, *Mathematical and Computer Modelling*, **37** (2003) 1143-1152.

M. Kolev, Mathematical modeling of the competition between acquired immunity and cancer, *International J. Applied Mathematics Computer Science*, **13** (2003) 289-296.

M. Kolev, E. Kozlowska, and M. Lachowicz, Mathematical model of tumor invasion along linear or tubular structures, *Mathematical Computer Modelling*, **41** (2005) 1083-1096.

N. Komarova, Spatial stochastic models for cancer initiation and progression, *Bulletin Mathematical Biology*, **68** (2006) 1573-1599.

N. Komarova and D. Wodarz, The optimal rate of chromosome loss for the inactivation of tumor suppressor in gene cancer, *Proceedings National Academy Science*, **101** (2004) 643-649.

N. Komarova, Stochastic modeling of loss- and gain-of-function mutation in cancer, *Mathematical Models and Methods in Applied Sciences*, **17** (2007) 1647-1674.

A.K. Kuemer, K. Tacheuchi, and M.P. Quinlan, Identification of genes, involved in epithelial-mesenchymal transition and tumor progression, *Oncogene*, **20** (2001) 6679-6688.

M. Lachowicz, Micro and meso scales of description corresponding to a model of tissue invasion by solid tumours, *Mathematical Models and Methods in Applied Sciences*, **15** (2005) 1667-1683.

L.D. Landau, E.M. Lifshitz, **Theory of Elasticity**, New York, Pergamon, (1970).

P.L. Lollini, S. Motta, and F. Pappalardo, Modeling tumor immunology, *Mathematical Models and Methods in Applied Sciences*, **16** (2006) 1091-1125.

S.A. Mani *et al.*, Mesenchyme forkhead 1 (FOXC2) plays a key role in methastasis and is associated with aggressive basal-like breast cancer, *Proceedings National Acaemy of Sciences*, **107** (2007) 10069-10074.

A.G. Marneros, J.E. Norris, S. Watanabe, E. Reichenberger, and B.R. Olsen, Genome scans provide evidence for keloid susceptibility Loci on chromosomes 2q23 and 7p11, *J Invest Dermatol*, **122** (2004) 1126-1132.

P. Martin and J. Lewis, Actin cables and epidermal movement in embryonic wound healing, *Nature*, **360** (1992) 179-183.

B.A. Mast, The skin, *Wound healing: Biochemical and clinical aspects*, (I.K. Cohen, R.F.

Diegelmann & W.J. Lindblad Eds), Philadelphia: Saunders, (1993) 344-355.

R.M. May, Uses and abuses of mathematics in biology, *Science*, **303** (2004) 790-793.

R.L. McCallion and M.W.J. Ferguson, Foetal wound healing and the development of antis-carring therapies for adult wound healing. In: **The Molecular and Cellular Biology of Wound Repair**, (Clark F.A.F. ed), 2nd edn., New York: Plenum Press, (1996) 561-600.

J.B. McCarthy, D.F. Sas, and L.T. Furcht, Mechanisms of parenchymal cell migration into wounds, in: **The molecular cell biology of wound repair**, R.A.F. Clark and P.M. Henson, Eds., Plenum, New York, **13** (1988).

F. Mollica, L. Preziosi, and K.R. Rajagopal Eds., **Modeling of Biological Materials**, Birkhäuser, Boston, (2007).

J.D. Murray, **Mathematical Biology**, III Edition, Springer, Berlin, (2002).

J.D. Murray and G.F. Oster, Cell traction models for generating pattern and form in mor-phogenesis, *J. Mathematical Biology*, **19** (1984) 265-279.

J.D. Murray, P.K. Maini, and R.T. Tranquillo, Mechanochemical models for generating biological pattern and form in development, *Physics Reports*, **171** (1988) 59-84.

G. Nadin, B. Perthame, and B. Ryzhik, Traveling waves for the Keller-Segel system with Fisherbirth term, *Interfaces and Free Boundaries*, **10** (2008) 517-538.

M. Naitoh et al., Gene expression in human keloids is altered from dermal to chondrocytic and osteogenic lineage, *Genes Cells*, **10** (2005) 1081-1091.

H. Nakaoka, S. Miyauchi, and Y. Miki, Proliferating activity of dermal fibroblasts in keloids and hypertrophic scars, *Acta Dermato-Venereol*, **75** (1995) 102-104.

F.B. Niessen, P.H. Spauwen, J. Schalkwijk, and M. Kon, On the nature of hypertrophic scars and keloids: a review, *Plast Reconstr Surg*, **104** (1999) 1435-1458.

D. Noble, Modeling the heart-from genes to cells to the whole organ, *Science* **295** (2002) 1678-1682.

M.A. Nowak and K. Sigmund, Evolutionary dynamics of biological games, *Science*, **303** (2004) 793-799.

G.F. Odland, Structure of the skin, **Physiology, Biochemistry and Molecular Biology of the Skin**, (L.A. Goldsmith Ed.), 3-62, New York, Oxford University Press, (1991).

L. Olsen, J.A. Sherratt, and P.K. Maini, A mechanochemical model for adult dermal wound contraction, *J. Biological Systems*, **3** (1995) 1021-1023.

L. Olsen, P.K. Maini, J.A. Sherratt, and B. Marchant, Simple modelling of extracellular ma-trix alignment in dermal wound healing I. Cell flux induced aligment, *J. Theoretical Medicine*, **1** (1998) 175-192.

L. Olsen, P.K. Maini, J.A. Sherratt, and J. Dallon, Mathematical modelling of anisotropy in fibrous connective tissue, *Mathematical Bioscences*, **158** (1999) 145-170.

L. Olsen, J.A. Sherratt, and P.K. Maini, A mathematical model for fibro-proliferative wound healing disorders, *Bulletin Mathematical Biology*, **58** (1996) 787-808.

H.G. Othmer, S.R. Dunbar, and W. Alt, Models of dispersal in biological systems, *J. Math-ematical Biology*, **26** (1988) 263-298.

H.G. Othmer and T. Hillen, The diffusion limit of transport equations II: Chemotaxis equa-tions, *SIAM J. Applied Mathematics*, **62** (2002) 1222-1250.

A. Palladini et al., In silico modelin and in vivo efficacy of cancer preventive vaccination, *Cancer Research*, **70** (2010) OF1-OF9.

B. Perthame and A. Dailbard, Existence of solutions of the hyperbolic Keller-Segel model,

Transactions of the American Mathematical Society, **361** (2009) 2319-2335.

M. Pinzani, Liver fibrosis, *Springer Seminars Immunophatology*, **21** (2000) 475-490.

O.J. Placid and V.L. Lewis, Immunologic associations of keloids, *Surg Gynecol Obstet*, **175** (1992) 185-193.

R. Reed, Why is mathematical biology so hard?, *Notices of the American Mathematical Society*, **51** (2004) 338-342.

P. Rosenau, Free energy functionals at the high gradient limit, *Physical Review A*, **41** (1990) 2227–2230.

P. Rosenau, Tempered diffusion: A transport process with propagating front and inertial delay, *Physical Review A*, **46** (1992) 7371-7374.

G.M. Saed, D. Ladin, J. Olson, X. Han, Z. Hou, and D. Fivenson, Analysis of p53 gene mutations in keloids using polymerase chain reaction-based single-strand conformational polymorphism and DNA sequencing, *Arch Dermatology*, **134** (1998) 936-967.

R.S. Schwartz, Shattuck lecture Ddiversity of the immune repertoire and immunoregulation, *The New England Journal of Medicine*, **348** (2003) 1017-1026.

F. Schweitzer, **Brownian Agents and Active Particles**, Springer, Berlin, (2003).

J.A. Sherratt, P. Martin, and J. Lewis, Mathematical models of wound healing in embryonic and adult epidermis, *IMA J. Mathematics Applied Medicine Biology*, **9** (1992) 177-196.

J.A. Sherratt and J.D. Murray, Models of epidermal wound healing, *Proceedings Royal Society London B*, **241** (1990) 29-36.

J.A. Sherratt and J.D. Murray, Epidermal wound healing: The clinical implications of a simple mathematical model, *Cellular Transplantation*, **1** (1992) 365-371.

I.I. Singer, D.W. Kawka, D.M. Kazais, and R.A.F. Clark, In vivo co-distribution of fibronectin and actin fibers in granulation tissue: Immunofluorescence and electron microscope studies of the fibronexus at the myofibroblast surface, *J. Cell. Biology*, **98** (1984) 2091-2106.

M.A. Stolarska, K.I.M. Yangjin, and H.G. Othmer, Multiscale models of cell and tissue dynamics, *Philosophical Transactions of the Royal Society A*, **367** (2009) 3525-3553.

D. Stopak and A.K. Harris, Connective tissue morphogenesis by fibroblasts traction I: Tissue culture observation, *Developmental Biology*, **90** (1982) 383-398.

Z. Szymanska, C.M. Rodrigo, M. Lachowicz, and M.A.J. Chaplain, Mathematical modeling of cancer invasion of tissues: The role of nonlocal interactions, *Mathematical Models and Methods in Applied Sciences*, **20** (2009) 257-281.

E. Tannenbaum and E.I. Shakhnovich, Semiconservative replication, genetic repair, and many-gened genomes: Extending the quasispecies paradigm to living systems, *Physics of Life Reviews*, **2** (2005) 290-317.

J. Tiuryn, D. Wójtowicz, and R. Rudnicky, A discrete model of small paralog families, *Mathematical Models and Methods in Applied Sciences*, **17** (2007) 933-956.

A.M. Turing, The chemical basis of morphogenesis, *Philosoophical Transactions Royal Society London*, **B237** (1952) 37-52.

B. Vogelstein and K.W. Kinzler, Cancer genes and the pathways they control, *Nature Medicine*, **10** (2004) 789-799.

Z. Wang and T. Hillen, Shock formation in a chemotaxis model, *Mathematical Methods in Applied Sciences*, **31** (2008) 45-70.

J.D. Watson, T.A. Baker, S.P. Bell, A. Gann, M. Levine, and R. Losick, **Molecular Biology**

of the Gene (5th ed.), Peason Benjamin Cummings (Cold Spring Harbor Laboratory Press), (2004).

G.F. Webb, **Theory of Nonlinear Age-Dependent Population Dynamics**, Marcel Dekker, New York, (1985).

R.A. Weinberg, **The Biology of Cancer**, Garland Sciences - Taylor and Francis, New York, (2007).

C.R. Woese, A new biology for a new century, *Microbiology and Molecular Biology Reviews*, **68** (2004) 173-186.

C. Xue and H.G. Othmer, Multiscale models of taxi-driven patterning of bacterial populations, *SIAM J. Applied Mathematics*, **70** (2009) 133-167.

J. Yang *et al.*, Twist, a master regulator of morphogenesis, plays an essential role in tumor methastasis, *Cell*, **117** (2004) 927-939.

Index